T0296143

OUR INTELLIGENT
BODIES

✽

OUR INTELLIGENT BODIES

GARY F. MERRILL

RUTGERS UNIVERSITY PRESS

NEW BRUNSWICK, CAMDEN, AND NEWARK,

NEW JERSEY, AND LONDON

Library of Congress Cataloging-in-Publication Data
Names: Merrill, Gary F., author.
Title: Our intelligent bodies / Gary F. Merrill.
Description: New Brunswick : Rutgers University Press, 2021.
Identifiers: LCCN 2019025250 | ISBN 9780813598529 (paperback) | ISBN
    9780813598512 (hardback) | ISBN 9780813598536 (epub) | ISBN
    9780813598543 (mobi) | ISBN 9780813598550 (pdf)
Subjects: LCSH: Human physiology. | Organs (Anatomy)
Classification: LCC QP34.5 .M486 2020 | DDC 612—dc23
LC record available at https://lccn.loc.gov/2019025250
A British Cataloging-in-Publication record for this book is available from the
British Library.

∞ The paper used in this publication meets the requirements of the American
National Standard for Information Sciences—Permanence of Paper for Printed
Library Materials, ANSI Z39.48-1992.

www.rutgersuniversitypress.org

Manufactured in the United States of America

CONTENTS

PREFACE

It is a sign of intelligence to love others. I love my wife, my mother, my maternal and paternal grandmothers, my mother-in-law, my daughters, my daughters-in-law, and my granddaughters. The last hours I spent with my mother were in a nursing home in June 2015. Even though she was diagnosed with advanced dementia, she recognized me, my wife, and our voices during those last hours. We hadn't seen her for a year, and she died peacefully fewer than 48 hours after our arrival. She was the last living member of her family of eight siblings; her oldest living sibling, a sister, preceded her in death by one week.

When our twin daughters were still at home and about age sixteen, their mother and they decided to shed a few pounds. To help achieve this, they transformed their eating habits to a more vegetarian-like diet. My wife warned me ahead of time. So by default, I joined them. After several weeks, I was "los'n weight and a look'n mighty frail" (I loved the voice and lyrics of Buck Owens and his songs). My wife and daughters were maintaining the status quo. Feeling sorry for me, my wife once again ramped up her baking and cooking, and I regained the pounds I had unwillingly shed. That was more than 20 years ago.

Now we also have five daughters-in-law. All appear to be in excellent health. They control their diets and weight and try encouraging their husbands and children to do something similar. Intelligently, our more disciplined sons look and act like their mother and respective wives. Some of my most stimulating intellectual conversations have been with these daughters-in-law. I regret not being physically closer to them to have such conversations more frequently.

My wife did not go as far in her formal education as our daughters and daughters-in-law. But when it comes to common sense and practical everyday living, I have never met her match. As much as I love the other women in my life, none are her equal. For example, because she is frugal, we have no debt. In the past 50-plus years, our only debts have been a home mortgage and my education. Because she is simple, our lives are not cluttered. Because she is service-oriented, the lives of others are more comfortable. If my wife and women like her ran the federal government and congress, there would be no U.S. debt. Regardless of the millions/billions of dollars they are worth, I doubt that any current or former U.S. chief executive or battalion of congresspeople can or could have managed a single dollar better than my wife does.

Contrary to what the above might sound like, this is not a book about the women in my life. Rather, it is a book about how our bodies display signs of intelligence at every level and in every way when left to their untampered-with physiological mechanisms. Intelligence can be seen in chemical reactions, molecular interactions, subcellular organizations, cell and tissue structures, and organ / organ systems functions. Even something as simple as changes in the volume and/or osmolality of body water compartments has potentially life-threatening consequences, yet our bodies show every sign of intelligence in readjusting to sustain homeostasis and life.

Another of my objectives in this writing is to convince the reader that intelligence is not limited to the central and peripheral nervous systems. As I argue in the first chapter, one definition of intelligence is the ability to solve problems. Our bodies are faced with challenging problems every day. For example, waking up after a good night's sleep finds us dehydrated and malnourished. Circulating concentrations of blood sugar are normally at their lowest levels. Upon waking, thirst mechanisms drive us to consume water whether we take it directly, indirectly, or intrinsically (dehydration problem solved). Hypoglycemic mechanisms encourage us to consume a morning meal: breakfast (malnourishment problem caused by a 12-or-more-hour fast is solved).

In the chapters that follow, I try to provide several examples of how our physical senses and organ systems display signs of intelligence in solving problems at every step. For example, the pigmented layer of the retina absorbs excess photons of light so they do not interfere with the physiology of photoreceptors and vision. When we lack adequate oxygen, the respiratory system adjusts by increasing both the frequency and volume of ventilation. Thus serious hypoxia is avoided and normoxia sustained in an intelligent display of negative feedback control.

I have intended to write to a college-educated mature adult audience (those at least having bachelor's or master's degrees plus considerable life experience). What I have written is not an invited critical review intended for those with PhD degrees (hence my purposeful lack of citations). For readers who want that level of detail, I suggest doing several PubMed searches using appropriate filters (e.g., reviews only; see https://www.ncbi.nlm.nih.gov/pubmed/). My final goal is to broaden the reader's thinking about the intelligence our marvelous bodies possess.

OUR INTELLIGENT
BODIES

1

INTELLIGENCE

INTELLIGENCE DEFINED

Intelligence has been defined as one's capacity for logic, abstract thought, understanding, self-awareness, communication, learning, emotional knowledge, memory, planning, creativity, and problem-solving. It is also the ability to perceive and/or to retain knowledge or information and to apply it. Intelligence is most widely studied in humans but has also been observed in nonhuman animals. Artificial intelligence is intelligence in machines and computer software (https://en.wikipedia.org/wiki/Intelligence, August 15, 2019).

The word *intelligence* derives from the Latin verb *intelligere*, "to comprehend" or "to perceive." A form of this verb, *intellectus*, became the medieval technical term for understanding and a translation for the Greek philosophical term *nous*. This term, however, was strongly linked to teleology and to the concepts of the active intellect (also known as the active intelligence) and the immortality of the soul. Such an approach to the study of nature was rejected by early modern philosophers including Francis Bacon, Thomas Hobbes, John Locke, and David Hume, all of whom preferred the word *understanding* in their English philosophical works. Hobbes, for example, in his Latin *De Corpore*, used *intellectus intelligit* (translated in the English version as "the understanding understandeth";

Goldstein, S., et al., eds., *Handbook of Intelligence: Evolutionary Theory, Historical Perspectives, and Current Concepts* [New York: Springer Science + Business Media, 2015]) as a typical example of a logical absurdity. The term *intelligence* has therefore become less common in English-language philosophy but was taken up later with the scholastic theories it now implies.

To some, the definition of intelligence is controversial. Indeed, when two dozen prominent experts were asked to define intelligence, they gave two dozen different definitions. Some psychologists have suggested the following definition: a general mental capability that among other things involves the ability to reason, plan, solve problems, think abstractly, comprehend complex ideas, learn quickly, and learn from experience. Intelligence is not merely book learning or test-taking smarts. Rather, intelligence reflects a broader and deeper capability to comprehend our surroundings—to "catch on," "make sense" of things, or "figure out" what to do.

Individuals differ from one another in their abilities to understand complex ideas, adapt effectively to the environment, learn from experience, engage in various forms of reasoning, and overcome obstacles by "taking thought." We have different intelligence quotients (IQs). One's IQ is computed based on the norm for his or her age group and circumstances. That norm is assigned a value of 100. Therefore, one's IQ can be either above or below 100. Although differences between individuals can be substantial, IQ is never entirely consistent among humans. One's intellectual performance will vary on different occasions, in different domains, and as judged by different criteria (http://www.unn.edu.ng/publications/files/ABULOKWE%20AMAECHI%20CLEMENT.pdf).

The widely varying human capacity to learn, store, and recall information can be illustrated by game shows such as *Jeopardy!*. Most contestants who qualify to be on the show are eliminated from the competition after a single round. Ken Jennings, however, holds the record for the longest winning streak on the show. He is also the second-highest-earning contestant in American game show history. In 2004, Jennings won 74 consecutive matches of *Jeopardy!* (nearly four months of appearances) before being defeated on his 75th appearance by challenger Nancy Zerg. Jenning's total earnings on *Jeopardy!* were $3,196,300, consisting of $2,520,700 over his 74 wins; a $2,000 second-place prize in his 75th appearance; a $500,000 second-place prize in *Jeopardy! Ultimate Tournament of Champions*; a $100,000 second-place prize in *Jeopardy! Battle of the Decades*; as well as half of a $300,000 prize in the IBM challenge, when he competed against Watson, the computer designed by IBM engineers.

THE OXFORD ENGLISH DICTIONARY AND INTELLIGENCE

According to the *Oxford English Dictionary* (www.oxforddictionaries .com), intelligence is the ability to acquire and apply knowledge and skills. Intelligence is also a person or being with the ability to acquire and apply knowledge and skills, according to *Oxford*. Further, intelligence is the collection of information of military or political value, and it is also the people employed in the collection of military or political information (*The Oxford American Dictionary and Language Guide*, New York: Oxford University Press, 1999). From the *Oxford English Dictionary Online* comes the following discussion of the U.S. Central Intelligence Agency and the gathering of covert intelligence: "the launching of the Central Intelligence Agency (CIA) on 18 September 1947 signaled an American addition to the customary use of the word *intelligence*." Referring to mental capacity, the word had carried one of two principal meanings. The first, archaic by 1947, simply indicated news. The second meaning covered information, at least partly clandestine and sometimes processed and analyzed, that might be of strategic importance (oed.com, August 15, 2019).

The advent of the CIA encouraged an additional meaning that had already been gathering pace and would solidify in the near future. The expanded definition came to embrace not just the gathering and cognitive processes but action as well. First in CIA parlance and then in general American usage, intelligence came to include covert operations, or the effort to influence politics in foreign countries by undercover means.

Deployment of the word *intelligence* was a way of making covert action more respectable. The battle was already half-won, as covert operations had come to be accepted and even admired in the Second World War, and the anxieties generated by the Cold War were predisposing people to accept peacetime practices that they might previously have questioned. But *intelligence* still had a better reputation than covert operations. It had come to be seen as a magic wand. There was a widespread belief that had U.S. intelligence not been in disarray, it could have prevented the Japanese attack on Pearl Harbor. Equally popular was the belief that improved intelligence had helped achieve victory in the naval battle of Midway and in the wider war. The word *intelligence* came to confer a respectability behind which the dirtiest of "dirty tricks" could hide (*The Oxford American Dictionary and Language Guide*, New York: Oxford University Press, 1999; accessed August 15, 2019).

<h2 style="text-align:center">MERRIAM-WEBSTER'S LEARNERS DICTIONARY AND INTELLIGENCE</h2>

According to Merriam-Webster (learnersdictionary.com), intelligence is "the ability to learn or to understand or to deal with new or trying situations; reason; also the skilled use of reason; the ability to apply knowledge to manipulate one's environment or to think abstractly as measured by objective criteria" (as tests; *Merriam-Webster's Collegiate Dictionary*, deluxe edition, 1998). The innate ability to "solve problems," "overcome obstacles," and "deal with trying situations" is among the definitions I will use to describe our intelligent bodies at the organ systems, organ, tissue, cell, subcellular, and even molecular levels. In arguing this way, I hope to persuade the reader that signs of intelligence as defined here and above are found ubiquitously throughout the human body. As a disclaimer, however, I do not think that each cell or subcellular organelle has its own little brain.

PHYSICAL AND PHYSIOLOGICAL CONCEPTS
THAT IMPLY INTELLIGENCE

Homeostasis—Walter B. Cannon (October 19, 1871–October 1, 1945) was an American physiologist who chaired the department of physiology at Harvard University circa 1906 to 1942. He was president of the American Physiological Society from 1914 to 1916. Cannon wrote the book *The Wisdom of the Body* that was first published in 1932. In his book and elsewhere, Cannon championed and popularized the idea of "physiological homeostasis" (extended from Claude Bernard's ideas of the constancy of the body's internal environment, i.e., the *milieu interieur*).

In 1915, Cannon coined the phrase "fight or flight" to describe an animal's response to threats. He outlined four tentative propositions to describe the general features of homeostasis:

1. Constancy in an open system, such as our bodies, requires mechanisms that act to maintain this constancy. Cannon based this proposition on insights into the ways by which

steady states—such as glucose concentrations (blood sugar),
body temperature, and acid-base balance—are regulated.

2. Steady-state conditions require that any tendency toward
change automatically meets with factors that resist change. An
increase in blood sugar (a potential problem) results in thirst
as the body attempts to dilute the concentration of sugar in the
extracellular fluid. It also results in the release of insulin as
the body helps the cells acquire access to elevated glucose.

3. The regulating system that determines homeostasis consists
of a number of cooperating mechanisms acting simultaneously or successively. Blood sugar is regulated by insulin,
glucagon, and other hormones that control its circulating
concentrations in the plasma, its release from the liver, and its
uptake by the tissues (the solution to the problem of excess or
insufficient circulating concentrations of blood sugar).

4. Homeostasis occurs not by chance but by the result of organized self-government and physiological intelligence.

Cannon's proposition number four above sounds like an application of intelligence as defined earlier. So the processes of elevations
in blood sugar (following ingestion of a meal), stimulation of thirst,
balanced and coordinated release of insulin from the pancreas, and
later release of glucagon from the liver are all part of an organized,
self-governing, intelligent system functioning at the molecular, subcellular, cellular, tissue, and organ-systems levels. The physiological
end results of such coordination are homeostasis of circulating concentrations of blood sugar, energy balance, and sustained health and
well-being of the body. Homeostatic regulation of blood sugar is
only one part of an endless list of physiological functions that intelligently sustain our physical bodies. Here is a short list of others I
will describe in later chapters: body water and its distribution, blood

flow and its regulation, body salt and electrolyte balance, and control of blood pressure.

According to Cannon, steady-state conditions require that any tendency toward change automatically meets with factors that resist change. A physiological process that is in the steady state is one that is not changing with time. Consider your own blood pressure. When you go for your annual medical examinations, you might be more or less excited. Perhaps you are anxious about being on time, being delayed in traffic, or missing work. You might park the car then walk briskly to the doctor's office. By the time the nurse escorts you into the examination room, wraps a pressure cuff around your upper arm, and listens for sounds as she deflates the cuff, your blood pressure is probably elevated. It would not be unusual for the nurse to call out numbers such as 135 over 85 or 140 over 90, especially in patients other than young adults, teens, and children. These numbers for systolic and diastolic pressures are on the modestly high side of normal. Even though elevated blood pressure is common with aging in older people, the above numbers can be nerve-racking during a hurried doctor's visit.

On occasions when I have been lying quietly for 15–20 minutes on an examination bed in my own laboratory, a research assistant (or other student being trained how to monitor blood pressure) can measure my blood pressure, and the numbers are about 120 over 80. These are noticeably lower than those mentioned above, and they fall within the physiological range for a healthy adult. The two sets of data were obtained under widely differing conditions. In the doctor's examination room, we are usually seated and often nervous about what lies ahead. Under such conditions, neither we nor our cardiovascular systems are in Cannon's "steady state."

Conversely, when lying quietly on my own examination bed, none of the above disturbances apply. If my research assistant measures my blood pressure immediately after instrumenting me, then again at 5-minute intervals during a 15–20 minute period, he will most likely find my pressures elevated on the first trial, a bit lower on the next, and not changing in the third and fourth measurements. In other words, after 15–20 minutes of resting quietly in the supine position, the physiological factors that regulate my blood pressure have reached Cannon's "steady-state" conditions, and my cardiovascular system is in a state of homeostasis.

Alternatively, if I have a 1-gallon container filled with water then punch a hole in the bottom of the container, the water will flow out at a rate determined by gravity and the size of the hole. The larger the hole, the greater the rate of flow, and the less time it will take for the water to escape. If I have a stopwatch and a graduated cylinder, I can collect some of the escaping water and calculate its rate of flow from the container. Imagine the hole is a small one and the rate of outflow is 100 milliliters per minute. After about 20 minutes at this rate, the container will have only about half its original volume (1 gallon is about equal to 4 liters or 4,000 milliliters). In 20 minutes more, the container will be empty. However, if I begin pouring water into the container when it is half empty, and if the rate of inflow (pouring) is 100 milliliters per minute, then the rates of inflow and outflow will be equal, and the container will remain half full for as long as the exercise is sustained. At such a point in time, the volume of water in the container is said to be "in equilibrium" because the two rates of flow are equal.

Now imagine that we substitute the 1-gallon container with a 1.25-gallon human circulatory system and the blood in it (about 5 liters). Imagine also that one's heart pumps those 1.25 gallons of blood around the entire system each minute. This rate of circulation, called the cardiac output, supplies the needs of all organs and tissues of the body with nutrients and oxygen each minute. The same circulating volume also removes waste products from the respiring cells and delivers them to excretory organs (e.g., carbon dioxide is exhausted by the lungs, and excess water and electrolytes are excreted by the kidneys). Under steady-state, homeostatic conditions, the rates of supply and demand for oxygen, as one example, are equal, and the body is in physiological harmony for oxygen. But what happens if we "punch a hole" in this circulatory container?

When blood is lost from the circulatory system faster than it can be replaced, the person is hemorrhaging. Hemorrhage leads to a reduced volume of circulating blood and to a state of hypovolemia (low blood volume). Hypovolemia causes a fall in blood pressure and a reduced supply of oxygen and nutrients to the tissues. Without the homeostatic mechanisms described by Cannon and others, this person could bleed to death. With homeostasis, the proper physiological adjustments will be made, and the person will survive.

Beyond Humans: Animal Intelligence

Solomon said about ants, "Go to the ant, thou sluggard; consider her ways and be wise" (Proverbs 6:6), and "The ants are a people not strong, yet they prepare their meat in the summer" (Proverbs 30:25). While thinking about animal intelligence, one biologist wrote that ants are so much like humans that they are an embarrassment to us. They farm fungi, raise aphids as livestock, launch armies to war, use chemical warfare (sprays) to alarm and confuse enemies, capture slaves, engage in child labor, and endlessly exchange information. The world of the ant has been intensely scrutinized in recent decades. And the notion that ants demonstrate signs of cognition has not been lost on these investigations. It is increasingly clear that some scientists have taken King Solomon's advice to heart.

Among ants, chemical communication can be compared to the human use of auditory, tactile, and visual communication. We shake hands to greet one another, and we give directions using fingers, arms, head nods, and other gestures. With ants, however, there is no cultural transmission of communication. For them, everything must be encoded in their genes. In humans, only basic instincts are carried in the genes of a newborn. Additional skills must be learned from other members of the community as the child grows and develops. It might seem that this cultural continuity gives us a great advantage over ants. However, ants' fungus-farming and aphid-herding crafts are sophisticated when compared to the agricultural skills of humans millennia or even centuries ago. The farming methods of ants have at least been sustainable since Solomon's time.

Intelligently, ants do not litter the environment with plastic bags, sandwich wrappers, plastic bottles, and cigarette butts. Moreover, recent evidence suggests that the crop farming of ants might be more sophisticated and adaptable than was originally thought. Ants were farmers millennia before humans were. Ants cannot digest the cellulose in leaves (a problem), but some fungi can. Therefore, ants purposely cultivate these fungi in their colonies and nests. The ants then bring the fungi leaves to feed on and use them (the fungi) as a source of food (problem solved).

Farmer ants secrete antibiotics to control other fungi that might act as weeds. They also spread waste to fertilize their crops. It was once thought that the fungus ants cultivate was a single species sustained from the distant past. But this is apparently not the case. Scientists recently screened the genetic makeup of 862 different subspecies of fungi taken from ants' nests and found them to be highly diverse. These scientists (myrmecologists) concluded that ants are continuously domesticating new species of fungi. Even more impressively, DNA analyses of the fungi suggest that ants improve or modify the fungi by regularly swapping and sharing strains with neighboring colonies.

Evidence also suggests that ants have lived in urban settings for millennia, developing and maintaining underground cities of specialized chambers, tunnels, and functions. When humans survey Mexico City, Tokyo, New York, or Beijing, they are amazed at what has been accomplished by other humans. Yet the magnificent books that have been published on ants since 1990 suggest that humans might also be outmatched by ants in their city- or colony-building ingenuity. For example, one pair of authors describe a supercolony of the ant species, *Formica yessensis*, on the Ishikari coast of Hokkaido, Japan. This megalopolis was reported to be composed of 360 million workers and a million queens living in 4,500 interconnected nests across a territory of 2.7 square kilometers.

Such enduring and intricately meshed levels of technical achievement outshine by far anything achieved by the distant ancestors of humans. We hail as masterpieces the cave paintings in southern France and elsewhere, which date back some 20,000 years. However, according to myrmecologists, ant societies similar to those of today existed more than 70 million years ago.

Other recent investigations have found evidence that ants can transmit technically complex messages. Scouting ants who have located food in a maze return to inform and mobilize foraging teams. As part of an experimental design, two groups of ants, scouts and foragers, were allowed to engage in contact sessions, at the end of which the scout was removed in order to observe what the foraging team would do without her. Commonly, the foragers navigated to the exact spot in the maze where the food had been found by the scout. These investigations have since focused on whether the route through the maze is communicated as a sequence of left and right turns or as messages of compass bearing and distances covered.

While walking to work one morning, I noticed a robin with a worm in his beak. I wondered how robins find worms. We have all noticed foraging robins hop several times, stop, cock their heads, then lunge at the ground. After driving their bills into the soil, they retrieve an earthworm and consume it or take it to feed their nestlings. I've guessed that robins either hear, see, or feel movements at the ground's surface. I have since learned that predatory birds (which robins are; among other living invertebrates, they also prey on earthworms) use smell, sight, sound, and possibly mechanical cues to find prey. One group of investigators conducted a controlled experiment in an aviary. In that study, four American robins found buried worms in the absence of sight, smell, and vibration. This suggested the use of sound to locate their prey. Their success in finding buried worms was significantly reduced by the presence of background white noise that obscured the birds' ability to hear.

I met my wife when she was riding her quarter horse, Mac Eb. This was at the Lincoln County Fair in Afton, Wyoming, August 1964. Marlene was an attractive teenage barrel racer. Her horse's name came from sire and dam, Mr. Eb and Scarlet Mac, both registered names among quarter horses of the time in Lincoln County. I've asked Marlene what signs of intelligence she observed in her father's quarter horses, including Mac Eb. She was taught to bring a bucket with a small amount of grain to the corral/pasture whenever she wanted to ride Mac Eb. Compared to hay, "grain is a cookie to them," Marlene said. Shaking a grain-containing bucket will either entice the horse to come or force him to run away. Both are signs of intelligence (yes, I want the cookie; no, I don't feel like being ridden today).

On June 6, 2015, American Pharoah (mistakenly misspelled by the owner, Ahmed Zayat of Teaneck, New Jersey) became the first horse to win the coveted Triple Crown in 37 years. The thoroughbred was just the twelfth horse to accomplish this feat in the 147-year history of the Belmont Stakes. Then on June 9, 2018, Mike Smith riding Justify won the trophy, making two such winners in just three years (the first time this had been done since the 1970s). Mike, at age 52, became the oldest jockey to win the Preakness Stakes and Triple Crown.

When the gates opened at the Belmont Stakes on June 6, 2015, Victor Espinoza took American Pharoah to the lead position on the rail within two to three strides and never relinquished the position or lead. About two minutes later and heading into the final turn, horse and jockey left the field of seven others behind as they sailed through the 1,097-yard stretch to the finish line in less than 25 seconds. During this historic race, American Pharoah wore makeshift earplugs. Both trainer and rider believed the horse was distracted by the noise of the crowds (e.g., Kentucky Derby, Preakness Stakes). By stuffing cotton in his ears before the race, they believed the horse was less likely to get distracted. Whether the earplugs helped is anyone's guess. However, that horses have intelligence can hardly be argued.

When measuring equine intelligence, some investigators begin with three assessments: (1) scope of learning, or the cognitive ability of a horse to solve increasingly complex problems; (2) rate of learning, a quantitative measure of the time required to learn a task; and (3) retention of learning, or the ability to remember the learned behavior. In addition, the Guide Horse Foundation has developed other measures of equine intelligence. These are the ear reflex, socialization, and Umveg tests:

Ear Reflex Index—All horses have a range of ear motion of approximately 170 degrees, and more intelligent horses demonstrate more frequent and independent ear motion. To administer this test in the field, one trainer stands near the left shoulder, and another trainer stands at the right flank of a horse. As the trainers walk around the horse, a more intelligent horse will follow the trainers with its ears, tracking the trainers' motions independently and rapidly from opposite ears. A less intelligent horse fixes its ears in the neutral position (facing 30 degrees forward).

Response to Socialization—As a herd animal, horses communicate frequently with other horses and learn a variety of social interactions. A more intelligent horse learns quickly to respond to social cues in a disciplined and masterful way. Horses also communicate in various ways, including vocalizations such as nickering, squealing, or whinnying; touching through mutual grooming or nuzzling; smelling and sniffing; and through body language. In addition, horses use a combination of ear position, neck and head height, movement, foot stomping, and tail swishing to communicate.

Two horses who have never met commonly approach each other with their necks extended and their heads bowed (unless they are wild mustang stallions competing for the same harem). They touch noses and exchange scents by blowing into each other's nostrils. Once the introduction is completed, they nicker, squeal, or whinny at each other. Horses are also concerned about their status in a hierarchy, and during this phase, they might challenge each other, sometimes striking out with their front legs. This behavior continues until one of the horses shows signs of submission.

Umveg Testing—This is the process of taking a detour in order to reach a goal. In horses, the ability to do an Umveg is an undisputed sign of superior intelligence. In one test, the horse is lead to a particular side of an open-ended, 10-foot wire fence. Next, the horse is lead to the other side of the fence and a treat is placed into a bowl directly opposite the horse. An intelligent horse will turn away from the food and circumnavigate the fence to get it, while a less intelligent horse will stand on the opposite side of the fence and paw the ground.

In a related test, a bucket with a small quantity of grain is presented to the horse. Then the grain is dumped onto the ground and the bucket is turned upside down covering the grain. A pole is placed, about 12–18 inches off the ground, between the horse and the bucket. A smart horse will walk around the pole, sniff the ground and bucket to confirm the grain is there, then push the bucket over with his nose and nibble at the grain. A horse of less intelligence will remain on the opposite side of the pole. Or he might step over the pole, usually knocking it off its stands, then sniff the ground/bucket before knocking the bucket over. These various responses are ranked and compared with other tests of intelligence to classify the horse as more or less intelligent. Because of their superior intelligence, some horses are being trained to replace guide dogs (guide horses: small breeds that don't require much space or large quantities of food and water).

Unintelligent Human Acts

Try to solve this problem: Jack is looking at Anne, but Anne is looking at George. Jack is married, but George is not. Is a married person looking at an unmarried person? Possible answers are yes, no, or cannot be determined. More than 80 percent of people answer this question incorrectly. If you concluded that the answer cannot be determined, you are one of them. The correct answer is yes, a married person is looking at an unmarried person. Most of us believe that we need to know if Anne is married to answer the question. But think about all the possibilities. If Anne is unmarried, then a married person (Jack) is looking at an unmarried person (Anne). If Anne is married, then a married person (Anne) is looking at an unmarried person (George). Either way, the answer is yes. Most people have the intelligence to figure this out if you tell them to carefully consider all the possibilities. But unprompted, they won't bring their full mental faculties to bear on the problem.

Failing to bring one's full mental faculties to bear on problems arguably contributes to many unintelligent human actions. Consider the images of a fallen global leader awaiting arraignment for alleged sexual molestation. Allegations that toppled Dominic Straus Kahn, former head of the International Monetary Fund, underscore a stunning capacity for unintelligent actions by otherwise intelligent people who should know better. Such unintelligent acts occur with numbing frequency. A U.S. Congressman seeks trysts over the internet; a prominent scientist fabricates data; millionaires cheat on taxes; a ranking general undermines the commander in chief; hapless investors turn over life savings to swindlers like Berny Madoff. And more recently, Alabama Governor Robert Bentley resigned in shame after mistakenly emailing love notes to his wife that were intended for his mistress and lover.

Journalist Jonah Lehrer rose to stardom by age 31. He wrote for prestigious publications like the *Wall Street Journal* and the *New Yorker*. In June of 2012, Lehrer got caught plagiarizing. Unlike most cases of plagiarism, though, Lehrer copied his own writing. He reused whole paragraphs from *Wall Street Journal* pieces that he wrote for the *New Yorker*. In a *New York Times* interview, Lehrer apologized for the plagiarism, saying it was just plain laziness that drove him to duplicate his work across the two publications.

That alone might not have been enough to permanently damage his career. After all, the words he copied were his own, even if it was unethical to use them without his publisher's knowledge. However, about a month later, the magazine *Tablet* accused Lehrer of making up Bob Dylan quotes for his book *Imagine: How Creativity Works* and then lying about it. *Tablet* reporter Michael C. Moynihan questioned Lehrer about some of the quotes in *Imagine*, and Lehrer told him that the quotes came from an old interview that had not been made public. Lehrer later admitted that he made up the quotes and that when Moynihan questioned their veracity, he panicked and lied about the source. When the story broke, *Imagine* publisher Houghton Mifflin Harcourt pulled the e-book and stopped all shipments of the book's physical copy. Lehrer resigned in disgrace as a writer for the *New Yorker*.

Dr. Andrew Wakefield published an article in the journal *Lancet* in 1998 that was based largely on fabricated data and falsifications. He had been paid $674,000 by lawyers hoping to sue the manufacturers of vaccines against measles, mumps, and rubella. Wakefield reported, falsely, to have found a link between autism and the vaccinations. The British General Medical Council revoked his license to practice medicine. The influence of Wakefield's lies and falsified data is still having an impact. As parents in the past several years have opted not to get their children vaccinated against these childhood diseases, a spike in cases of measles has been reported.

President Bill Clinton apologized to the American public and to the world on December 11, 1998, only after it was incontrovertible that he had extramarital sexual relationships with staff member Monica Lewinsky. Initially, Clinton vehemently denied having any sexual relations with Lewinsky, but as the evidence against him became insurmountable, he had to admit his immoral and unintelligent actions. CIA Director General David Petraeus also initially denied his extramarital sexual affair with his biographer Paula Broadwell. His final admission led to his resignation and shame as he fell from grace.

Judgment, if not plain common sense, flees at immense cost, and consequences destroy reputations, careers, and freedom (some lead to prison terms). Catastrophic failures of judgment share common threads including, but not limited to, inflated egos and unchecked arrogance. Sadly, the same unrestrained drive that pushes people to succeed can also push them to catastrophe.

In summary, any two or more people can argue the definitions of intelligence. They can also debate the existence of both animal and plant intelligence. Here I have tried to include the ability to identify and solve problems as a sign of intelligence. If you agree with this definition, then you probably also agree that ants and horses, among a long list of other animals, display signs of intelligence and the ability to solve problems.

2

PHYSIOLOGY OF LIGHT
AND VISION

Some years ago, I started the practice of walking part of the distance to work each morning. I decided to time my morning/evening routines with the rising and setting of the sun. To watch these indescribably celestial events, one's dedication and determination must correspond with the seasons of the year and the timing of the event. For example, if a sunrise occurs at 5:30 a.m., I have to be up by 4:00–4:30 a.m. to shower, shave, exercise, eat breakfast, travel to a designated parking lot, and walk to the site of my observations. Watching a sunrise is like attending a concert. The main event is usually preceded by an introductory act. To take it all in, you must be in your seat on time.

A sunrise can last 3 or 4 minutes—or 45–60 minutes depending on your perspective and definitions. If you are there just for the 3- or 4-minute main event, your timing must be impeccable. If you want to take in the pre- and/or postshow events, timing is less important. Sixty minutes before the sun rises, the eastern horizon is dark and colorless. After another 15–20 minutes, signs of light appear. Simultaneously from north to south, the horizon displays increasingly brilliant shades of campfire-orange light. These are distributed equidistantly in either direction. In another 15–20 minutes, they begin to converge near the point of sunrise. As convergence reaches its crescendo, the first diamond-like shimmerings of the sun appear through the trees (for me in Central New Jersey). Within the next 3–4 minutes, the sun will have fully risen. The entire show is indescribable, and momentarily, I am filled with deep gratitude for life; my standing ovation.

No two sunrises are completely alike. Each is a singular event. Among other places, I have watched them in the woods and on the Atlantic shores of New Jersey, on the plains of Western Nebraska, at 38,000 feet over the Pacific Ocean, and across the valley from the Grand Tetons (Jackson Hole). They are all inspiring and more than amazing. Music in the classic Broadway play *Fiddler on the Roof* contains the following lyrics: "Is this the little girl I carried? Is this the little boy at play? I don't remember growing older, when did they? Sunrise, sunset . . . swiftly flow the years . . . one season following another, laden with happiness and tears." I love the play and the music, but I cherish more the memories of little boys and little girls growing older.

The Physics and Physiology of Light

Light from the sun combined with other materials causes a form of energy called electromagnetism. Like the waves of the ocean, light waves travel over distances and have heights (amplitudes). The distance between two successive electromagnetic waves is its measurable wavelength. The height of each wave is its zenith, and the trough is the nadir (or peaks and valleys). The frequency is the periodicity of the waves. The length of electromagnetic light waves can be expressed on a light spectrum. The portion of the spectrum that is discernible to the human eye is called the visible light spectrum. The visible light spectrum is about three- to four-tenths of 1 micrometer in length. A micrometer is indiscernible to the unaided human eye, and it takes 1,000 micrometers to equal one millimeter (clearly discernible to the unaided eye). One thousand millimeters equal one meter (about 39 inches).

Some electromagnetic waves other than visible light have wavelengths one-thousandth (e.g., soft x-rays) to one-millionth of a millimeter (e.g., gamma rays). Others have wavelengths that are meters (e.g., microwaves) or even hundreds of meters long (e.g., radio waves). At the shortest end of the electromagnetic spectrum, light looks dark blue or purple and is called *ultraviolet*. At the longest end of the electromagnetic spectrum, it looks dark red and is called *infrared*. Visible light of different wavelengths exists between these two end points. It ranges from shades of violet and blue at one end to orange and red at the other.

Traveling from any source, light passes through various media before it reaches the retina of the eye. One of these media is air. Surrounding the earth is an ionosphere, a space of charged chemical particles called ions. Ions carry either negative or positive charges. Also, as light nears the earth, it is influenced by magnetic fields at both poles. Thus light reaching our eyes from the sun or from any other source travels as electromagnetic waves influenced in part by ions and magnetism. Any medium (environment) through which light passes can influence both its path and its velocity. Light is absorbed, reflected, refracted, stored, and transmitted by these media. One reason the daytime sky appears to be blue is that all other electromagnetic waves get absorbed by particulate matter as light passes through space (i.e., the reds, oranges, yellows, and greens).

If light waves hit a solid object that does not absorb them, they get reflected back to their source (e.g., the sun). If light waves hit an object at an angle, they get refracted or bent so that their path and velocity are changed, but light is not returned to the source. Any medium that allows light to pass through it is said to transmit light. The atmosphere, pure water, and similar media are effective transmitters of light because they neither alter the course of light nor reduce its velocity. Other media absorb light and convert it into chemical and electrical energy.

In one of our laboratory experiments, we teach students about urine and the composition of water and solutes in it. To demonstrate the ability of the kidneys to either concentrate or dilute urine, we have students measure their urine in various states of body hydration. They do this using refractometers. Refractometers are instruments used to estimate the concentration of solutes in urine (or other solutions and suspensions). Refractometers influence the passage of light through a urine sample placed on a prism (part of the refractometer). Concentrated urine refracts the light pathway and slows its velocity more than dilute urine. By this methodology, students can tell if they are dehydrated or overhydrated.

How Does Light Affect Vision?

If light gets refracted as it comes in contact with an object or medium, the degree of refraction can be determined. Light that passes through a vacuum with no particulate matter has a refractive index (or index of refraction) of 1.0003. This means the vacuum alters neither the velocity nor the angle of light as it passes through it. Pure water—that is, water that has had the particulate matter removed (including ions)—also has a refractive index near 1.0. Thus the refractive index of a medium is an expression of the angle and speed of light within it.

Once light leaves the atmosphere and comes in contact with the eye, it must pass through at least four media of differing refractive indexes. Only then does it reach the retina, where an image is formed and the information is relayed to the brain. In order, the four media are (1) the cornea, (2) the aqueous humor behind the cornea but in front of the lens, (3) the lens, and (4) the vitreous humor (the gelatinous substance that fills the eye between the lens and the retina). Also in order, the refractive indexes of these four media are 1.376, 1.336, 1.386, and 1.336 (remember, air has an index of refraction of 1.0003).

The cornea has no blood vessels, and its tremendous transparency is caused by specialized epithelial cells. The pupil is the port of entry of light into the eye. The iris surrounds the pupil. The degree of neuromuscular-generated contraction/relaxation of the iris determines the volume of light passing through the pupil. Also, the iris is the colored structure that can be seen through the translucent cornea. The color of the iris comes from pigments in its cells. More pigment produces darker hues (black, brown eyes), and less pigment causes lighter hues (blues, greens, yellow-green, etc.).

The ciliary structures (body, epithelium, muscles) attach to the iris and help regulate the flow of light through the pupil. The pupil is like the aperture of a camera, and the iris is analogous to the diaphragm. Together they determine the volume of light allowed to enter the camera/eye. When an optometrist or ophthalmologist shines a flashlight beam into your eye, he is trying to observe whether the "pupillary light reflex" is normal. When you are being examined by an ophthalmologist for potential eye disease, he first drips medication on your eye. The medication blocks neuromuscular reflexes, so the pupil does not constrict in response to light. By doing this, the ophthalmologist can pass more intense beams of light into the eye for longer periods of time. This allows him to inspect the retina in detail that could not be achieved if the pupils were constricted.

After leaving the ophthalmologist's office on such occasions, especially in brilliant sunlight, I must wear dark glasses and wait several hours before safely driving my car. Alternatively, nowadays, most ophthalmologists require the patient to have someone else drive him home while the pupillary light reflex is still impaired. Control of pupillary size in both eyes is coupled. That is, if the doctor shines a light in just one eye, its pupil will constrict (the so-called direct light response). If one's physiological optic reflexes are intact, the shining light will also cause an identical constriction in the opposite eye (the so-called indirect or consensual light response).

Epithelial cells of the ciliary body continuously secrete a protein-free aqueous fluid that is an ultrafiltrate of blood plasma. Ultrafiltrate means it contains all the elements of plasma except proteins (e.g., water, electrolytes, ions, and other inorganic and organic chemicals). This fluid, or aqueous humor, fills chambers behind (posterior) and in front of (anterior) the iris. By drainage and production being balanced, the volumes of fluid in each chamber remain relatively constant in the healthy eye. They also cause a modest pressure of about 20 mmHg inside the eye and help give the eye its overall shape. If the canals that drain the anterior chamber are blocked, the volume of aqueous humor increases and can cause glaucoma or an elevation in pressure. Such elevated pressure compresses delicate sensory nerves that carry information from the retina to the optic nerve. It can also compress blood vessels and impede blood flow. Too much compression for too long will permanently damage the eye and cause loss of vision.

The lens is an onion-like structure with concentric layers of densely packed epithelial cells that are arranged in columns. These are encased in a tough, thin, and transparent capsule. The cells of the lens have a high concentration of the protein α-crystallin, which helps increase the density and focusing power of the lens. The lens is also a convex structure on both sides, so it has the power to bend the stream of light flowing through the pupil. This helps focus light near the central fovea of the retina. Important as the lens is, it is not as important as the cornea, which does most of the initial bending of light as it enters the eye. Moreover, the shape of the lens can be made more or less elliptical or round depending on the strength of contraction of the ciliary muscle attached to the iris.

As photons of light enter the retina, they pass through several layers of neuronal cells within it. Sequentially, these are the ganglion cell layer, inner nuclear layer, inner plexiform layer, outer nuclear layer, outer plexiform layer, and the photoreceptor layer (rods and cones). The outermost layer (farthest from the vitreous humor) is the pigmented epithelium. These are cells that absorb excess light and perform a host of other functions. Once an image has been formed on the retina, sensory information about the image is returned to the optic nerves through these same cell layers only in reverse sequential order. Then via the optic nerves, the image and its information are transferred to the central nervous system for processing.

Because the photoreceptors are the last to receive photons of light, some have questioned whether this was a flaw in design. However, others have argued that there was no mistake. Embryologically and histologically, the retina seems to be a displaced structure of the central nervous system. The rods and cones are immediately adjacent to the pigmented cell layer that overlies the vascular blood supply. Compared to the much thinner and transparent neuronal layers of the retina, the epithelial pigment cells and underlying choriocapillaris (ocular blood vessels) are opaque and poorly transparent. Photoreceptors are extremely delicate and undergo continuous damage, repair, and renewal. Close proximity to the nutrients and gases of the ocular blood supply is critical to these processes. Moreover, excess photons are absorbed by the pigmented epithelial cell layer, thus preventing them from otherwise distorting the resolution of an image being processed and transferred to the optic nerve. The photoreceptor layer could be in no other retinal location and still perform its intended functions.

FUNCTIONS OF RODS AND CONES

Of the structural components that aid in vision, the most important is the retina. There are an estimated 15–20 times more rods than cones in the human retina. Rods are cylindrical and cones are conical. The human eye has only one kind of rod, and it is responsible for dark-adapted or monochromatic vision (one color; also called scotopic vision). There are three types of cones, and they are responsible for the color vision (photopic vision) we experience in daylight or brighter environments. Rods and cones are distributed throughout the retina in a distinct pattern.

In the center of the retina is a tiny depression called the fovea or central fovea. The characteristics of the central fovea are markedly different from those of adjacent areas of the retina. Light that is at the center of our gaze or field of vision is concentrated and focused in the fovea. The fovea has only cones and no rods. Therefore, the ratio of photoreceptors to other neurons is considerably reduced compared to the retina at large. Because of this, foveal, or central, vision has the highest resolution and the greatest acuity. In other words, the light-receptive field in the fovea is small for each photoreceptor compared with other areas of the retina.

Because foveal vision is mediated entirely by cones, each layer of photoreceptors consists of only the smallest cones packed at the highest densities (i.e., the shortest distances from the center of one cone to the center of the next). Outside the fovea, the density of cones falls as that of rods increases. Therefore, peripheral vision (i.e., vision other than foveal) is mediated by both rods and cones. At the periphery of the retina, the ratio of photoreceptors to ganglion cells is high, thus each unit of cells (photoreceptor, ganglion cell, bipolar cell) has a large receptive field. This reduces the spatial resolution in the region but increases its sensitivity because there are more photoreceptors per unit space to collect light.

It is helpful to use the terms *proximal* and *distal* as landmarks for a discussion of the path light follows as it passes from the anterior chamber and lens through the vitreous humor and retina. Thus as light enters and passes through the retina en route to the photoreceptors, it flows in a proximal to distal direction. As light continues toward the pigmented layer underlying the photoreceptor layer, it still moves in a proximal to distal direction.

Once a photon reaches either a rod or a cone, it passes sequentially through three distinctive regions of the cell. From proximal to distal, these are the following:

1. The synaptic terminals, a region of the photoreceptor that communicates with other neuronal cell types.
2. The inner segment of the photoreceptor is composed of the nucleus and cell body and an adjacent region where the metabolic machinery, including mitochondria, is found. This region synthesizes photopigments.
3. The outer segment is where the physiological stimulus and response are mediated. Some biologists refer to this region of the photoreceptor as the so-called transduction site. The outer segment of the rod or cone is the site where light, as an electromagnetic form of energy, is converted to a physiological response that can be conveyed back to the central nervous system.

Photochemistry of Dark and Light Vision

Since about 1940, research on vision has led to some marvelous discoveries. For example, as early as 1942, we learned that it takes only a few photons, each acting on a single rod, for the human eye to detect light as a signal. This demonstrated that the eye is operating at the edge of its physical limitations; there is no known measure of light smaller than a photon. To detect a single photon requires tremendous amplification of a light signal. One investigator said that if the energy needed to lift a cube of sugar one centimeter could be converted to the energy of the blue-green portion of the electromagnetic spectrum, it would produce a burst of light so intense that the eyes of every person who ever lived could detect it.

Once inside the outer segment of the photoreceptor, a beam of light comes in contact with hundreds of layers of stacked disks that are completely separated from the cell membrane (rods) or are continuations, but not separate subcellular organelles, of the outer membrane (cones). Membranes of disks in rods contain the photopigment rhodopsin. In cones, the membrane-like disks contain similar molecules that are related to rhodopsin. Rhodopsin is transported in small vesicles from its site of synthesis in the inner segment of the photoreceptors to its site of action in the outer segment of the same rod.

Changing light energy to the electrical and chemical events needed to convey a signal to the brain requires detection, amplification, and conduction. Photoreceptors, like other sensory receptors (e.g., sound, taste), are the detectors. They use electrical events to conduct and transmit light signals. Photoreceptors do this by changing membrane electrical potentials. Photoreceptors are activated when photons of light hyperpolarize their resting membrane potentials. In most other sensory receptors, activation is caused by depolarizing resting membrane potentials. Thus light causes the cell membrane potentials of rods and cones to become more negative than the resting potentials that are maintained in the dark.

Membrane hyperpolarization is an essential step in relaying visual signals. Hyperpolarization increases the rate of release of neurotransmitters, and neurotransmitters are necessary for the conduction of light signals through the various layers of the retina. Synapses between adjacent layers of retinal neurons are conventional in that more transmitter is released during depolarization and less is released during hyperpolarization. Thus a flash of light reduces the release of neurotransmitters. One end result is that human rods are most active in the dark.

Investigators of the physiology of light and vision have found that both the outer and inner segments of photoreceptors contain ion channels in their membranes. Membrane currents passing through these channels are carried by calcium, potassium, and sodium ions. Potassium moves out of the cells in the inner segments, while sodium and calcium move into the cells in the outer segments. In the dark, when cones are not activated and rods are functioning at their peak activities, outward movement of potassium and inward movement of sodium produce a resting transmembrane potential of about −40 millivolts (mV).

Experimentally, when a single rod is isolated and a minimally intense flash of light is passed through it, the resting membrane potential of about −40 mV becomes slightly more negative (e.g., about −41 or −42 mV). Responses to flashes of light are graded and become more negative as the intensity of the flash increases. The most intense flash of light will hyperpolarize the resting membrane potential to about −60 or −65 mV. Greater intensities are saturating—that is, they reach a plateau where no further change in membrane potential is produced regardless of the strength of the light stimulus.

Flashes of light cause sodium channels in rods to close. This prevents sodium from entering the cell. At the same time, light has no influence on potassium channels of the inner segment. Therefore, potassium channels remain open as light passes through the rod. This means that there is a net loss of positive charge (via potassium efflux and blocked sodium influx) from the cell when it is exposed to light and that the net loss of charge causes the membrane to hyperpolarize (i.e., to become more negative on the inside of the cell).

The number of sodium channels that are closed by light depends on the number of photons that are absorbed. The range of sensitivity of a single rod is about 1–1,000 photons. It has been estimated that a single photon, by blocking sodium channels, can prevent the flow of one million ions into the photoreceptor. This illustrates the tremendous amplification of photo energy in the human eye. Cones are less sensitive to light, but their ion channels open and close more rapidly, and they do not reach a plateau as light intensity increases.

The physiological mechanism behind light vision begins with the so-called light receptor molecule, rhodopsin. *Rhodon* comes from Greek for the color "rose," and the English translates *rhodon* to "red." The process of converting dark or night vision to light or day vision begins with rhodopsin. Rhodopsin is a protein, and among all physiologically important proteins, rhodopsin is about the most tightly packed. The packing of a protein refers to the number of molecules that can occupy a specific unit of space on a sensory receptor (e.g., one square micrometer on a rod disk). Quantitatively, there are as many as 1,000 disks per rod in the human eye. Each disk packs up to 30,000 rhodopsin molecules per square micrometer of the membrane, and a single rod has been estimated to contain at least one billion rhodopsin molecules. This almost unbelievable density of proteins optimizes the photon-absorbing potential as light passes through each rod. Even so, human rods use only about 10 percent of the light they are exposed to. The rest is either captured by other optical components of the eye or is absorbed by the retinal pigment epithelium that overlays the retinal photoreceptors.

Rhodopsin consists of two subunits: retinal, a form of vitamin A or retinol, and opsin, a much larger molecule. For light signals to be processed (transduced), photons must be absorbed by retinal (which gives rhodopsin its red color). Absorption of light by retinal changes its molecular conformation. One conformation of retinal is unstable and exists only in the dark. The other conformation is more stable and can exist in the light. Once retinal absorbs a photon, the conformational change occurs within a picosecond (one-trillionth of a second). This initial light-induced change in retinal leads to a secondary conformational change in opsin and to a subsequent cascade of molecular events.

In order for the photon stimulus at the disk membrane to be converted to an electrical signal in the membrane of the photoreceptor, a sequence of chemical events in both is required. In the dark, when retinal is attached to the much larger protein opsin, retinal has a kinked tail, leaving it in the 11-cis retinal configuration. In this molecular shape, 11-cis retinal sits in a pocket of opsin and is covalently bound to it. The 11-cis retinal confirmation gives rhodopsin its red color and is an unstable molecule.

When a photon of light is absorbed by rhodopsin, the 11-cis retinal is transformed into a more stable molecule called all-trans retinal. The process of transformation is called isomerization. Isomerization of retinal causes corresponding changes in the molecular configuration of opsin, leading to a form called metarhodopsin II. Soon after these conformational changes, the newly created all-trans retinal and metarhodopsin II separate in a reaction called "bleaching," which changes the color of rhodopsin from the bright red of well-oxygenated arterial blood to the lighter yellowish hue of urine or plasma (opsin's color). The photoreceptor cell then converts all-trans retinal to retinol (vitamin A), which makes its way to the retinal pigment epithelium. There, retinol is reconverted to 11-cis retinal and is subsequently transported back to the outer segment of the photoreceptor, where it recombines with opsin. This cycle (the conversion of all-trans retinal back to 11-cis retinal) takes several minutes for completion.

The preceding changes in chemistry alter the structure of another molecule that is attached to metarhodopsin II, and it is called transducin. It was given this name because it transduces the light signal detected at the disk membrane into an ionic/electrical signal at the outer membrane of the photoreceptor. Transducin could also have been called "convertin" for its ability to convert a light stimulus into a physiological response (membrane hyperpolarization) at the outer membrane. Be that as it may, transducin was discovered in the mid-1980s as the first recognized member of the superfamily of molecules that bind guanosine triphosphate (GTP) and have come to be known as G-binding proteins (or simply, G proteins).

The synthesis and breakdown of G proteins are accomplished by enzymes and signals that lead to a balance in second messengers called cyclic nucleotides. One of these is cyclic guanosine monophosphate (cyclic GMP, cGMP). Cyclic GMP is the second messenger that couples the photon-activated events of the disk membrane to the electrical events of the outer membrane of the rod. Another key discovery in the mid-1980s revealed that the light-sensitive ion channel of rods is actually a cGMP-operated cation channel.

In the dark, the enzyme that synthesizes cGMP (guanylyl cyclase) keeps the concentration of cGMP high inside the rod. This causes the cGMP-operated channels to remain open much of the time. In turn, this establishes a so-called dark current, which is nothing more than the near-continuous movement of sodium ions (and some calcium ions) into the outer portion of photoreceptor coupled with the efflux of potassium out of the inner segment of the protein. This flow of current into and out of the cell leaves the cells membrane potential near −40 mV. The enzyme that breaks down cGMP is called phosphodiesterase (PDE). It is sensitive to light and stimulated by it. Because a photon of light stimulates PDE, the concentration of cGMP inside the cell is reduced. This leads to the closure of cGMP-operated cation channels, thus reducing the dark current. The photoreceptor membrane is hyperpolarized (becomes more negative), the release of neurotransmitters from its presynaptic terminals dwindles, and a light-induced visual signal is created and transferred to more proximal retinal neurons.

For some readers, the above might not be sufficient detail. For others, it might be too much. To the former, I would recommend a PubMed search using key phrases such as "physiological reviews" and "chemistry of night vision." Then to narrow the search, use PubMed filters such as "reviews," "free full text," "humans," and "2015–2020." These will narrow the field, and you should find a small handful of very helpful review articles written in recent years that summarize the field in detail. To those for whom the above is too much, I apologize.

FUNCTIONS OF THE PIGMENTED RETINAL EPITHELIUM

The pigmented retinal epithelium increases the quality of the optics of light by forming a dark barrier separating retinal photoreceptors from the retina's underlying blood supply, which is called the choriocapillaris circulation. This barrier assists in the absorption of light that is not captured by rods and cones. The outer retina is also exposed to an oxygen-rich environment. Blood perfusion of the choriocapillaris averages about 1,400 milliliters per minute per 100 grams of tissue. This is much higher than most of the generously perfused tissues in the body, including the gut, brain, heart, peripheral chemoreceptors, and kidneys (25–50, 40–60, 50–75, 100–150, and 300–400 milliliters per minute per 100 grams of tissue, respectively). Venous blood from the choriocapillaris shows a 90 percent oxygen saturation compared with 98 percent for general arterial blood and 25–75 percent for venous blood in other organs (e.g., heart at 25 percent, inactive skeletal muscle at 75 percent). This suggests minimal extraction of oxygen during the passage of well-oxygenated arterial blood through the choriocapillaris.

On the other hand, venous blood from retinal vessels has an oxygen saturation of 45 percent. This comparison suggests considerable differences in the metabolic activity of tissues of the eye. The retina is thought to float on the choriocapillaris, which appears to function as a bed of blood-filled vessels. Unfortunately, such an arrangement is also ideal for promotion of light-induced oxidative damage to the retina. The proximity of pigmented epithelium and blood flow is exacerbated by the continuous production of reactive oxygen species during phagocytosis to shed outer segments of the photoreceptors. Optimistically, however, such juxtaposition also means ready access of the retina to a generous venous blood supply to wash away reactive oxygen species and other injurious by-products of metabolism within the adjacent tissues.

Other than the venous washout just mentioned, the retinal pigmented epithelium has three lines of defense against light-induced oxidative damage. The first line is absorption and filtering of light. The retinal pigmented epithelium possesses a complex variety of pigments that correspond to different wavelengths and the wavelength-dependent risks of damage that light bears. General absorption of light occurs via the skin-pigmenting hormone melanin in melanosomes of the retina. This is supplemented by additional absorption by photoreceptors.

Photoreceptors contain as their most important pigments lutein and zeaxanthin (carotenoids). These pigments act as physiological sunscreen that absorbs blue light. Blue light appears to be most dangerous for retinal pigment epithelium in the adult eye because it promotes the oxidation of lipofuscin to cytotoxins. It has been reported that 10 milligrams of lutein plus two milligrams of zeaxanthin per day reduces the risk of age-related macular degeneration but does not prevent the disease. Conversely, exposure of the adult retina to ultraviolet light is low because the lens absorbs most of the UV light. However, melanosomes and blue-light-absorbing pigments are only responsible for the absorption of 60 percent of light energy. This implies the presence of other retinal pigments that have not been described yet. One of these could be lipofuscin because it accumulates in the retinal pigment epithelium during life.

The second line of defense consists of naturally occurring antioxidants. The retinal pigment epithelium contains high amounts of superoxide dismutase and catalase, naturally occurring enzymatic antioxidants. As nonenzymatic antioxidants, this tissue accumulates carotenoids (such as lutein and zeaxanthin), ascorbic acid (vitamin C), alpha-tocopherol (vitamin E), and beta-carotene (precursor to vitamin A or retinol). These are supplemented by glutathione and melanin, other naturally occurring antioxidants. The cell's physiological ability to repair damaged, biologically important macromolecules such as DNA, lipids, and proteins is its third line of defense against oxidative damage.

Age-related macular degeneration is the most common cause of blindness in people living in industrialized nations. Also, the accumulation of advanced glycation end-products (AGEs) and their receptors (RAGEs) increases in the retinal pigmented epithelium with age. One of the most effective actions against both is adequate daily consumption of foods containing vitamins A, C, and E. Also, as we age, an imbalance between the cytoprotective actions of the retinal pigment epithelium and the toxic effects caused by oxidative damage grows. This too contributes to retinal degeneration.

Increased transport of products between adjacent epithelial cells within the retinal pigment epithelium helps control the accumulation of fluid (retinal edema), and decreased epithelial transport contributes to retinal degeneration. In the visual/retinal cycle, various forms of retinol are isomerized and transferred continuously between the outer segments of photoreceptors and the retinal pigment epithelium. Inherited mutations in the genes that operate and maintain this cycle contribute significantly, with age, to retinal degeneration and loss of vision. So too does the failed ability of the pigmented epithelium to phagocytose worn out photoreceptors.

The retinal pigmented epithelium also buffers ions, including acids and bases, in the subretinal space (microscopic region between the outer segments of the photoreceptors and the retinal pigmented epithelium). Such buffering maintains homeostasis of acid/base balance in these tissues and ensures the physiological excitability of photoreceptors. Finally, the retinal pigmented epithelium secretes a variety of growth factors, and age-related failure of this function contributes to proliferative retinal disease.

COLOR VISION

Color vision is sometimes referred to as photopic or light vision. Three different sets of photoreceptor cones are used to detect and process color vision. These are red cones (toward the infrared end of the spectrum), blue cones (toward the ultraviolet end of the spectrum), and green (or yellow; central region of the electromagnetic light spectrum). Our eyes respond to only a small region of the electromagnetic light spectrum—that is, to light wavelengths about 400 nanometers, or 0.4 of a micrometer, in length. In daylight, we see an assortment of colors because objects absorb light of some wavelengths while reflecting, refracting, scattering, and transmitting light of other wavelengths. A scene's color can also be affected by light of different sources. For example, tungsten lamps emit light that is various shades of red. Fluorescent bulbs emit a bluish light, and halogen, neon, and fluorescent lamps can produce superbright shades of amber, blue, red, and white light. These gas-containing lamps, as well as light-emitting diodes (LEDs), are used increasingly in the headlamps and taillights of new cars and emergency vehicles.

The trichromatic theory of vision (three-colors theory) was first enunciated by Thomas Young in about 1800. Herman von Helmholtz helped champion the theory later in the 19th century. By mixing the proper intensities of three colors of light (blue, red, and green; i.e., the primary colors), these investigators found that they could reproduce the hues of various experimental samples of light. They first proposed that color vision, and our ability to perceive a wide range of different hues of color, is based on three pigments in the eye, each with the ability to absorb light of differing wavelengths. It would take about another 150 years for instrumentation and science to advance sufficiently to prove the trichromatic theory. Microspectrophotometry and the ability to isolate and study individual rods and cones did not occur until the mid-1960s.

Moreover, our sensitivity to discriminate color vision depends on the degree of adaptation of the retina to light. The eye is either dark-adapted or light-adapted. That is, we see objects, shapes, shades of gray (darker or lighter), and color depending on the state of adaptation of retinal rods and cones to differing intensities of light. Cones can adapt to low intensities of light relatively quickly (e.g., in less than 5 minutes). Conversely, rods adapt to dimmer light more slowly, taking as many as 25–30 minutes to adjust.

When the retina is dark-adapted, our vision adjusts to shorter wavelengths of light (toward the ultraviolet end of the spectrum; i.e., scotopic vision). When the retina is adapted to light, it adjusts to wavelengths at the longer end of the spectrum (toward the infrared end; i.e., photopic vision). The maximum sensitivity to light can be several orders of magnitude higher under scotopic than photopic conditions. The main explanation for these differences is that rods are sensing the intensity of light under dark-adapted conditions, and cones are transducing light intensity under light-adapted conditions.

The spectral sensitivity of the light-adapted eye depends on the photopigments in the cones, and that of the dark-adapted eye depends on rhodopsin in the rods. The three populations of cones are called blue, green and red historically because the peaks of light waves they absorb occur at the violet (or blue; short wavelengths), green–yellow (medium wavelengths), and yellow–red (longer wavelengths) ends of the electromagnetic spectrum. Today, to avoid confusion and disputation, investigators simply refer to cones as S, M, and L for short, medium, and long wavelengths, respectively.

MELANOPSIN AND INTRINSICALLY PHOTOSENSITIVE RETINAL GANGLION CELLS

In 1998, investigators searched cDNA libraries of frog skin melanophores to find molecules whose amino acid sequences closely resembled rhodopsin and violet opsin. They reasoned that light-sensitive photopigments in frog skin would bear some resemblance to those from the eyes of frogs and other vertebrates. They did indeed find a photopigment with about 30 percent homology to vertebrate opsins,

and in situ hybridization revealed its expression in the melanophores of frog skin.

Like other visual pigments, this novel photopigment, which
they named melanopsin, had a predicted structure with seven
regions of the molecule passing through the membrane of the
chromophore (color-producing cell). Oddly, further analysis of
the similarities in amino acid sequences revealed that melanopsin
bore a greater similarity to invertebrate opsins than to the typical molecules found in vertebrate rods and cones. Also surprising
was the fact that these frog melanopsin precursors were found in
tissues other than the skin. For example, they were found in brain
structures such as the suprachiasmatic nucleus and in eye components such as the iris, the retinal pigment epithelium, and the
inner retina. This led to speculation and later demonstration that
melanopsin is involved in regulating circadian rhythms, sleep patterns, and the pupillary light reflex.

The above turn-of-the-21st-century discovery led to more recent
investigations in mice that were genetically engineered to eliminate
rods and cones. The so-called "rodless" and "coneless" mice were
found to still be responsive to bursts of light. This finding suggested
the presence of photoreceptors other than rods and cones. Mice are
nocturnal. If just prior to initiation of the light-induced phase of
their sleep cycle such mice were kept in the dark, the onset of sleep
could be delayed. Conversely, if mice were exposed to flashes of bright
light during the dark phase of their routine daily activity, onset of
sleep could be hastened. This shift in phase of light/dark activity was
found to be due to the more recently discovered intrinsically photosensitive retinal ganglion cells (ipRGC) and their ability to produce
melanopsin. The extent of the shift in phase is influenced by the
intensity of light, which reaches a plateau. In melanopsin knock-
out mice (those lacking the genes to manufacture melanopsin), the

plateau to light intensity is markedly reduced when compared with melanopsin wildtype mice (those with the genes intact).

Today, we know that the ipRGC cells play an important role in the pupillary light reflex, circadian photoentrainment, and sleep regulation. We also know that other vertebrates, and even plants, have photoreceptors in places other than the eye. As research in this field continues, we undoubtedly will discover photosensitive tissues other than retinal rods, cones, and ipRGC even in the human.

LUNAR CYCLES

At about the same time as my experiences with sunrises, I was also treated to repeat observations of lunar cycles. To me, lunar cycles are no less amazing than sunrises and sunsets. On one occasion, I was able to watch the rising/setting full moon and rising/setting sun all within a 12- to 14-hour period. I remember the event because it occurred after the death of a dear friend. Occasionally, one can see the slimmest waning/waxing crescent moon in close proximity to a major planet (e.g., Venus). And if you are determined and the weather permits, during several sequential nights/early mornings, one can observe the changing proximity of the moon and a planet and/or of two or more planets. Among the phases of the moon, my favorites are the crescents. I think this is because the slimmest crescent moons do not obscure my view of the late-night or early morning skies.

I have always been struck by the similarities one can imagine between the lunar cycle and a woman's reproductive cycle. The lunar cycle is divisible into quarters according to the shape of the moon. Beginning with the new moon, there are the waxing crescent, waxing gibbous, waning gibbous, and waning crescent subcycles. The cycle from one new moon to the next can also be divided into four weekly phases. Like a complete lunar cycle that takes about 28 days, the woman's monthly reproductive cycle lasts an average of 28 days. Her 28-day cycle, arguably, can be divided into four phases. These are the menstrual, ovarian, luteal, and uterine phases. In humans and other mammals, each of these phases has been well studied and thoroughly documented. The first 14 days of the lunar cycle leads to the development of a "full moon." The first 14 days of a woman's reproductive cycle creates a mature ovarian follicle (a spherical object reminiscent of the spherical full moon). After this, a luteal phase lasts about 24 hours. It leads to the eruption of the mature follicle, the release of an ovum (egg), and the loss of the spherical ovarian follicle. This is analogous to the early waning gibbous phase of the moon when the spherical shape is lost. There are other similarities, but they are beyond the scope of my musings here.

The lunar and solar cycles remind me that there are only three natural measures of time. The solar day is marked by the rising and setting of the sun (measure number one). The lunar cycle (measure number two) is defined by the return of the "new moon." And the 365-day calendar year is quantified by the return of the seasons (the autumnal year; measure number three). All other expressions and measures of time are man made and dependent on the above three. Thankfully, man has devised units of time such as the hour, the minute, the second, and the nanosecond. After all, where would the Olympic swimming, ice skating, and track events be without the millisecond? And how would international economies, trade, and travel survive without the hour and minute (e.g., opening and closing of stock markets, intercontinental oceanic shipping)?

My wife and I were reminded of the oceanic shipping lanes and time zones (man made largely because of international shipping) while we were at the Dwight D. Eisenhower Lock near Massena, New York. In a few hours, we were able to watch large cargo ships pass from the Atlantic Ocean into the St. Lawrence Seaway or vice versa after they had been elevated/lowered 30 or 40 feet, respectively, in that lock. This was an educational experience, and temporally, it fell right on the heels of a recent boat ride we had taken in the Thousand Islands region of the St. Lawrence River near Astoria Bay, New York.

Concluding Thoughts on Light and Darkness

If the way our eyes and minds detect and process light can be associated with intelligence, then voluntarily and sometimes ignorantly descending into the abyss might be argued as evidence of lack of intelligence. In recent years, behavioral investigators have kept rats and other experimental animals in the dark for hours, days, and even weeks. After six weeks of light deprivation, rats display signs and symptoms of depression. Autopsies reveal that neurons in areas of the rat brain that are associated with cognition, emotion, and pleasure begin to atrophy. Neurotransmitters released by these neurons, such as dopamine, norepinephrine, and serotonin, are also reduced by prolonged exposure to darkness.

Cavers (spelunkers) are a group of adventure-seeking men and women who take pleasure in exploring the deep and dark recesses of caves, abandoned mines, and other subterranean environments. These explorers are not the faint of heart, and they must have absolute freedom from claustrophobia. Indeed, it is these very characteristics, among others, that often lead them into harm's way and put the lives of others at risk. Consider the case of Jason Storie and Andrew Munoz and their near-ill-fated adventure into Cascade Cave on Vancouver Island, December 2015 (see various versions and videos online).

The average yearly precipitation on Vancouver Island can exceed 250 inches at Henderson Lake on the west coast (the wettest place in

North America). Unpredictably heavy rainstorms can occur at any time of the year. The heaviest rainfall takes place in the fall and winter months. Apparently, without checking a weather forecast, Storie and Munoz and a few others descended about 150 feet below the surface into an abandoned mine without even knowing that it had begun to rain. The heavy rain quickly began filling the labyrinth of tunnels, channels, and waterfalls they were passing through. While exiting, Storie got stuck in an hourglass-shaped constricted tunnel. His lodged body, acting like a cork, helped fill the water spaces upstream, and he was in danger of drowning.

Munoz returned to assist Storie, and they were eventually able to dislodge him. However, both men were forced to spend nearly 14 hours in the cold, dark catacomb until the storm and water subsided the next morning. By that time, nearly 50 rescuers from several agencies had converged on the site to assist. Both cavers spoke later of their thoughts of death and dying in that dark and unforgiving graveyard. Munoz's first child had been born only six months earlier, and he wondered if his little girl would grow up fatherless. Storie spoke with a God he wasn't sure he believed in and with his parents, who had died earlier in his life. Had they paid closer attention to the weather forecasts, perhaps the incident could have been avoided.

3

HEARING AND SOUND

PHYSIOLOGY OF HEARING AND SOUND

As youths, my friends and I experimented with the travel of sound by yelling at granite walls in the nearby canyons of the Rocky Mountains. The sounds seemed to bounce from rock face to rock face as they traveled ever-farther from us and became increasingly muffled. We also experimented with the flashes of lightning and the delayed sounds of thunder thinking we could judge the distance between us and the lightning. None of us knew anything about the physics or physiology of sound and hearing. We did not know what organs and tissues lie beyond the eardrum (tympanic membrane) nor the detailed structures and functions of the middle and inner ears. We also gave no thought to how aging, illness, and/or injury could harm hearing. Now, as older men, our appreciation for these things has grown significantly as we more frequently say "Huh?" during conversations. If you are a younger man or woman reading this book, please take better care of your ears and hearing now. Be more selective about what you listen to and the volumes at which you listen, including using earphones with music. Be still and listen! Good communication begins with good listening.

What Is Sound?

Like electromagnetic waves that carry light to our eyes, so too is sound brought to our ears by waves. The space around us is filled with atmospheric air. Sounds alter atmospheric pressure, and pressure gradients move air in the form of sound pressure waves. Consider the sounds of a cannon or train whistle. Inside the Rutgers University football stadium, whenever the football team scores a touchdown, cannoneers clothed in Revolutionary War attire fire a colonial cannon. When field goals are scored, a freight train whistle sound is made. This has been a tradition for as long as I have been on the faculty at Rutgers University. The tradition might have begun with the first football games between Rutgers and Princeton, circa 1869, but I cannot confirm this.

Whenever the cannon is fired, atmospheric air immediately surrounding it is displaced and set in motion. Adjacent masses of air are compressed and decompressed (rarefied), leading to the formation of sound-induced pressure airwaves. These airwaves rapidly pass through the atmosphere of the stadium and its surrounding environs. If I am seated 75–100 yards away from the cannon, I see the puff of smoke and feel the compressed air milliseconds before I hear the sound. If I am close to the cannon, the smoke and sound occur nearly simultaneously. This observation demonstrates that sound travels at measurable velocities (350–400 meters per second). Even though the cannon's noise is not good for my hearing, I love the tradition and excitement that accompany the touchdown. The signal marks progress by the football team. But more importantly, the cannon's sound and its operators remind me of patriots who helped make my life of comfort and modern conveniences possible because of their selfless service and sacrifice. Thank you, men and women, patriots of the American Revolution.

The levels of such sounds have been defined and are quantifiable. As we all know, just because we don't hear a sound does not mean it is not there (e.g., the sensitivity of a bat's or whale's hearing vs. our own, the deaf's inability to detect the sound). The compressed/decompressed air inside Rutgers football stadium creates a measurable sound pressure level (SPL, or Lp). If one wants to quantify the SPL, he must have the proper equipment and use conventional units of expression. These units include Pascals (Pa) or microPascals (μPa, one-millionth of a Pascal) and decibels (dB, or one-tenth of a bel, named in honor of the inventor of the telephone, Alexander Graham Bell). The Pascal was named after Blaise Pascal, French mathematician, physicist, inventor, and writer (ca. 1623–1662). Pascal's laws apply to the transmission of pressure in aerodynamic, hemodynamic, and hydrodynamic systems. Of course, nowadays, one can estimate the SPL caused by the firing of the Rutgers University cannon by simply taking out his smartphone, opening the *Sound Meter* application, and holding the phone in the air when the cannon is fired.

The scale for expressing SPL in Pascals/microPascals is large and cumbersome. The scale for decibels is not. Thus when reporting sounds for consideration by the general public, it is customary to convert Pascals to decibels. This can be done using logarithms and the SPL. One SPL = 20 Log_{10} (P/P_0), where SPL = sound pressure level, Log_{10}, = logarithmic scale to the base 10, P = the actual measured sound, and P_0, = the threshold sound or the lowest/softest SPL detectable by the healthy young adult ear. P_0 is considered to be 20 μPa. As a frame of reference, the following familiar sounds and their corresponding SPLs, expressed in Pascals and decibels, are (1) a quiet room, 0.002 Pa or 40 dB (HVAC in my office is about 40 dB); (2) a quiet conversation at 3 feet, 0.02 Pa or 60 dB; (3) a diesel truck and its engine at 50 feet traveling about 40 mph, 0.5 Pa or 85 dB; and (4) a rock 'n' roll band, 15 Pa or 110 dB.

I like the voice of Steve Perry and some of the music of the band
Journey. Some years ago, I took my wife to Trenton, New Jersey, to
see and hear the band. We had cheap seats on an upper level of the
venue near the convergence of the ceiling and walls. The band's volume was so loud that I felt my heart palpitating and my eardrums
about to burst. We chose to leave after a couple numbers. In stark
contrast to Journey's sounds is the complete absence of noise inside
the anechoic chamber (lack of echo) of the Orfield Laboratory in
Minneapolis, Minnesota, arguably the quietest place on earth (as
certified by Guinness). The chamber is used, among other things,
to test audio equipment. It has double walls made of concrete and
reinforced steel 12 inches thick. Inside these walls, the room is lined
with 36-inch-thick shelves and wedges of soundproofing fiberglass
acoustic material. When the two heavy, vault-style doors are closed
and the lights turned off, sound inside the chamber plummets to
-9 dB, the quietest sounds (or lack thereof) ever recorded. Most
subjects tested in the anechoic chamber of the Orfield Laboratory
last 20 minutes or fewer. That level of silence is mentally and emotionally disturbing. Visitors have never experienced it before. Experimental subjects report hearing and feeling only the sounds of their
bodies (e.g., the borborygmus noises of the gastrointestinal tract,
whistling-like breathing, and heartbeats—i.e., most likely the pulsa-
tility of blood flow perfusing the inner ear).

I was an adult leader camping with a group of 12- to 16-year-old
boys some years ago. We were in the woods of northwestern New
Jersey. I was returning to my tent one evening in the dark. On the
path in front of me were three of the youngest boys. I could tell
from their conversation that they were inexperienced and from the
Newark, New Jersey, area. They did not know I was following them
and eavesdropping. The woods were absolutely silent and still. One
boy said to the others, "Don't you just hate it out here?" The others

replied, "Yeah!" The dark woods and palpable silence were unchartered territories for them. Their comfort zones were already established by the crowded neighborhoods and frequent sirens of the Newark environs. I felt bad for them.

When estimating one's blood pressure, medicine has relied on sound for many decades. Consider your last or a recent physician's medical examination. Before the doctor sees you, a nurse, medical assistant, or other trained person measures some vital signs. These include body temperature, body weight, heart rate, and blood pressure. Blood pressures, both systolic and diastolic, are estimated using a pressure cuff and a stethoscope (for detecting sound). The nurse wraps the pressure cuff around an upper arm. She inflates the cuff to a pressure of 180–200 mmHg. Then she places the bell or diaphragm of her stethoscope over the brachial artery in the same arm that is cuffed. As she deflates the cuff, blood flow that had been occluded when pressure in the cuff was 180–200 mmHg is restored.

As cuff pressure drops to about 120 mmHg and the occluded artery is partially opened only during the systolic phase of the cardiac cycle, small volumes of blood are pumped through the partially occluded artery. These small volumes cause blood vessel walls and surrounding soft tissue to shake and vibrate. The vibrating sounds are transmitted to the stethoscope, where they are amplified and used to estimate systolic blood pressure. As the cuff pressure declines below about 80 mmHg, the sounds are no longer heard. At this point of silence, the patient's diastolic blood pressure is estimated to be about 80 mmHg, and blood flow is fully restored throughout the entire cardiac cycle.

In one of my teaching laboratory experiments, we use devices other than stethoscopes to detect arterial blood pressure, and we ask the students if these alternate devices are suitable substitutes for the standardly used stethoscope. One such device is a cardiomicrophone.

It is a small circular transducer with an external diameter of about 15 millimeters, a thickness of 8 millimeters, and a circumference modestly greater than a nickel. The device has a built-in microphone. The cardiomicrophone is taped over the brachial artery just downstream of the lower edge of a blood pressure cuff. This sound-sensitive transducer has a cable leading from it. The cable is attached to an analog-to-digital (AD) converter data acquisition system. The AD data acquisition system is coupled to a desktop computer running software used to analyze sound and other physiological variables. The cuff is pressurized and depressurized just as with a stethoscope, and the sounds detected at systolic pressure are amplified, filtered, and recorded as sound waves on a visible computer monitor. When deflation of the cuff achieves diastolic pressure, the amplitudes of the sound waveforms become uniform (compared to the increasing amplitudes detected at systolic pressure). Students conclude that the cardiomicrophone can be used reliably to record one's blood pressure.

General Structures of the Human Ear

The mammalian ear, including outer, middle, and inner components, is a marvelous set of organs. The ear can only be fully appreciated by a detailed study of both its structure (anatomy) and function (physiology). For the interested reader, I recommend learning about structure from Dr. Frank H. Netter and his 13-book series called the Netter Collection of Medical Illustrations. This series includes the greater fraction of some 20,000 of his paintings that have become one of the most famous medical works ever published. For the reader who is short on time, the *Netter Atlas of Human Anatomy* was published in 1989. Its 25th anniversary edition (sixth edition, 2014) is a good place to begin for a more concise summary of Netter's illustrations.

Pages 94–100 of *Netter's Atlas of Human Anatomy* present a systematic overview of the structure of the human ear. This overview begins with topics and illustrations such as the "Pathway of Sound Reception," "External Ear and Tympanic Cavity," "Bony and Membranous Labyrinths," and so on. As you review these paintings and develop an interest, you can then perform PubMed searches for the corresponding physiology and, eventually, pathophysiology.

To study the physiology of the mammalian/human ear, I recommend consulting several classic and more recent editions of *Physiological Reviews*. Alternatively, go to the internet and find PubMed (www.ncbi.nlm.nih.gov). In the search box, type *hearing* or any related keyword or short phrase. This could yield more than 100,000 articles. To fine-tune your search, add filters such as the following: "Reviews Only" (reduces the number of publications to under 10,000), "Free Full Text" (reduces the search to 1,500 articles or so), "Pub Dates" (e.g., last five years, custom dates; reduces publications to several hundred), then finally "Humans Only" (additional reduction). To narrow the search even further, use more specific terms (inner hair cells, outer hair cells, osseous semicircular canals, etc.).

Structurally, the outer ear was designed to complement the environment and conditions the animal lives and survives in. Some species of mammals have unusually large auricles, pinna, or external ears. Consider whitetail deer and your most recent encounter with them on the trail. Other mammals have extremely small external ears, such as the Arctic fox. His ears are diminutive to reduce surface area and the loss of vital body heat to the extremely cold environment he lives in. Many mammals that are either predators or potential prey can move and reposition their external ears to help focus and concentrate on sound from a particular area.

The Middle Ear and Eardrum

The famous Carlsbad Caverns and the structures of the middle and inner ears are not unlike. The Carlsbad Caverns are composed of multiple compartments and rooms. They cannot all be seen in a single day, and several require the visitor to be part of a ranger-guided tour. There are also self-guided tours, the main entrance fee, and additional entrance fees for the ranger-guided tours. In the summer of 2014, my wife and I visited this world-famous national park. We camped in a nearby state park (daytime high temperatures in the area that July were about 105–115°F).

At the cave, one has the option of two modes of entrance. The first is a self-guided, nearly 750-foot descent below the earth's surface (1.5 linear miles because of multiple switchbacks, turns and uphill/downhill slopes in the paved trail). The second is a faster elevator ride. Not wanting to miss anything, we chose the 1.5-mile self-guided entrance into this auditory canal–like environment. As one descends, it becomes progressively darker and cooler. After leaving ambient temperatures from the 90s to the low 100s, the cooldown in the first few minutes is most welcome. The daytime average temperatures in the deepest recesses of the cavern are about 55°F. Therefore, if you plan to spend several hours in the cave, carrying a sweater or light jacket is an intelligent idea. The total self-guided tour of an estimated 4–5 miles will take 1.0–1.5 hours to descend/ascend (each), plus another 2–3 hours sight-seeing, eating lunch, and visiting the gift shop. You will probably need the sweater or jacket.

As one tours the cave, at least three things other than the amazing formations will be evident. First is the changing light and darkness. At times, the darkness is welcome. At other times, increasing light is absolutely needed. The second thing you will notice is the rapid decline in temperature. This might take more or less time depending on one's thermoregulation. And third, the absence of sound might be disturbing to some. The only things you will hear are the sounds of footsteps, someone's whispers to you or to someone nearby, and occasional shuffling of bodies as visitors pass one another at crowded points in the trail.

Our intent was to take only part of a day to descend the cavern, see the Great Room, and return before dusk to watch the bats exit en masse for their evening feeding. Therefore, we did not see the following rooms: the Hall of the White Giant, King's Palace, Spider Cave, Left Hand Tunnel, or Slaughter Canyon Cave. Each one of these rooms is seen only via ranger-guided tours and might take several days. Some of them require low-crawling through tight passageways, like Matlock's Pinch, as well as freestyle and ladder climbing. The benefits of staying multiple days to see more rooms include learning their histories of discovery and seeing formations such as the White Giant, the Monarch, and the Queen's Draperies (and others like soda straws, mushrooms, and helictites).

Like the main entrance to Carlsbad Caverns that collects and accumulates visitors, the pinnae of the ears collect sound and sound waves. As the self-guided tour pathway delivers visitors to the deep and deepest recesses of the cavern, so does the external ear's auditory canal deliver sound and sound waves to the eardrum and to the middle and inner ears.

Beyond the eardrums lie the middle and inner ears, each with its various contents and purposes. These chambers are analogous to the different rooms and contents inside Carlsbad Caverns. The middle ear is separated from the auditory canal by the eardrum. It is filled with air (as opposed to fluids). The middle ear consists of the epitympanic recess and its formations, including three tiny bones: the hammer, anvil, and stirrup (the smallest bones of the body). They are connected to each other, and the stirrup is attached to the oval window of the cochlea. A tissue membrane covering the oval window separates the middle ear from one of the channels of the coiled cochlea. Another membrane covering the round window also separates the middle ear from the cochlea.

Beneath the epitympanic recess and its contents is the tympanic cavity. The tympanic cavity widens to form the mouth of the Eustachian tube. In adults, the Eustachian tube is longer than in children, and it runs obliquely to the auditory canal (where it is horizontal to the canal in children). At its distal end, the Eustachian tube opens into the nasopharynx. The close anatomical proximity of the Eustachian tube and the middle/inner ear chambers helps explain how sinus infections, colds, and sore throats can be transferred from the mouth, nose, and sinuses into the middle and inner ears.

INNER EAR: SEMICIRCULAR CANALS, VESTIBULE, AND COCHLEA

Immediately downstream from the middle ear is the membranous labyrinth, composed of semicircular canals, the vestibule (with utricle and saccule), and the cochlea. Collectively, these organs and their lymph-like viscous fluids (endolymph and perilymph) compose the inner ear, a mostly fluid-filled chamber. There are three semicircular canals on each side of the head. They are oriented in three different planes and are filled with endolymph and perilymph. Because they are bilateral organs and are oriented in different planes, the semicircular canals can detect movements of the head in any imaginable position (the pitch, roll, and yaw axes). Their design is to monitor the head's (bodies) three-dimensional orientation in space, its angular acceleration and deceleration, and gravitational influences. The horizontal canals detect angular acceleration when the head is rotated to the right or left (as in shaking the head *no* to a question).

The anterior vertical canals detect movements of the head in the anterior/posterior directions (as in shaking the head *yes* to a question or when drawing the chin to or away from the chest). The posterior vertical canals detect lateral movements, such as when the ear is moved toward the shoulder.

The vestibule is the central component of the membranous labyrinth of the inner ear. It lies in close medial proximity to the ossicles of the middle ear, upstream to the cochlea, and adjacent to the semicircular canals. The vestibule contains two important organs, the utricle and saccule. They recognize linear acceleration of the head/body and help determine the body's orientation in space (e.g., standing upright vs. on one's head, lying down vs. reclining in a chair). The word *vestibule* derives from the Latin *vestibulum*, and literally means "entrance hall." At its junction with the cochlea, the vestibule has two windows, one oval (above) and one round (below), that communicate with different channels within the cochlea.

The first home I remember living in was a small, two-bedroom, one-level, wood-frame house built by my father in western Wyoming about 50 miles south of Jackson Hole. The front door opened into a modestly expanded space my parents called "the vestibule" (my first introduction to the word and concept). In the vestibule were a throw rug, wall-mounted coat hangers, and space for muddy boots and shoes. When visitors entered our house, the sounds of their voices could always be heard coming from the vestibule. I recall with fondness cheerful greetings as aunts and uncles, cousins and friends, and nearby grandparents entered our vestibule on cold winter nights in the 1950s and 1960s.

In the inner ear and downstream of the vestibule and its contents is the cochlea. In the adult, the cochlea is a coiled, snail-shaped organ about the size of a pea. It is the hearing apparatus of the human ear. Its base (proximal end) opens into the vestibule, and its apex (distal end) is found about two and one-half coils downstream (about 35 millimeters in the adult). Looking into the base of the cochlea (like looking into a garden hose), one can see that the tubular structure is divided into three separate channels. From top to bottom, they are the scala vestibuli, scala media, and scala tympani. *Scala* derives from the concept and words for circular staircases. The three scalae are separated by membranes. Reissner's membrane separates the scala vestibuli from the scala media. The basilar (or basal) membrane separates the scala media from the scala tympani. The moving parts of the cochlea, including hair cells, exceed one million in number, making it the most complex mechanical structure in the human body.

From the base of the cochlea to its apex, the basal membrane is characterized, among other features, by four rows of specialized hair cells (three at the outer edge of the basilar membrane and one toward the inner edge). Similar hair cells are found in the semicircular canals, where they open into the vestibule, and inside the utricle and saccule of the vestibule. All hair cells in structures of the inner ear are sensory. At their apex, hair cells have tiny finger-like structures called stereocilia. Stereocilia (sometimes called microvilli) project into the lymph-like fluids that fill the cochlear scalae. Similar stereocilia are found at the junctions of the semicircular canals and utricle and inside the saccule. Some stereocilia extend into open extracellular space, whereas others project into fluids/gelatinous matrices of their host structures.

In the utricle and saccule, tiny calcium carbonate crystals, embedded in a gelatinous matrix, come in contact with the stereocilia of the hair cells. As our heads/bodies move, so do the lymph-like fluids, the calcium carbonate crystals (called otoliths), and the stereocilia. Also, as sound pressure waves travel through the inner ears, the basilar membranes (on which the hair cells are positioned) are set in motion during both compression and decompression phases of the sound pressure waves. A compression phase bends the stereocilia in one direction, and the decompression phase bends them in the opposite direction.

The portion of the basilar membrane that contains hair cells is called the organ of Corti. At the organ of Corti, the stereocilia of the outer hair cells come in contact with the tectorial membrane above them. Stereocilia of the inner hair cells are free-floating in the endolymph of the scala media. The organ of Corti forms a sensory epithelium where hair cells are found in close proximity to supporting cells (see Netter illustrations). A thickened area of proliferating cells, the prosensory domain, appears in the floor of the cochlear duct during development. This region gives rise to the cochlear sensory epithelium. The reason loss of hearing is permanent, it is thought, is because hair cells do not regenerate after they are lost (e.g., due to aging, disease, medication, surgery, accidents).

As sound waves cause changes in pressures and volumes of the cochlear channels, they move the basal membrane and its hair cells. This movement causes stereocilia to bump against the tectorial membrane, where they are bent in one direction or the other, thereby stimulating (or inhibiting) sensory nerves at the base of hair cells. Electrical signals (action potentials) created by movements of the hair cells and their stereocilia are conducted through sensory nerves to the brainstem, the thalamus, and other regions of the brain. These signals inform the central nervous system of the sounds we hear (like the welcome vestibule voices from my childhood) and the movements we feel. The basilar membrane and its hair cells are more sensitive to high-frequency sounds at the basal end of the cochlea and to low-frequency sounds at the apical end. Also, by the time sound pressure waves reach the oval window, they have been amplified more than 20 times, revealing the amplifying power of the middle ear.

Naturally, the pressurized endolymph/perilymph and their movements are subject to the influences of gravity and altitude—hence the so-called popping of the ears detected when driving from lower to higher elevations or when liftoff occurs in a commercial airplane. As we rise in elevation, we pass through a continuum of declining ambient pressures—that is, the pressure surrounding us (atmospheric or barometric) gets lower the higher we go. This creates detectable pressure gradients between the endolymph of the semicircular canals and the atmosphere. As the pressure outside the eardrum decreases and that of the inner ear remains near-constant, the difference across the eardrum causes it to move, and we sense this as a popping sound or sensation.

Endolymph-induced latency (its movement lagging behind that of the moving head) is the cause, in part, of an illusion in pilots called "the leans." As a pilot guides his plane into a turn, hair cells in the semicircular canals are stimulated by movements of the endolymph. Stimulation tells the brain that the aircraft and its pilot are no longer moving in a straight line. If the turn is sustained indefinitely, the flow of endolymph eventually matches the movement of the ducts and ceases to stimulate the stereocilia. At that point, the pilot no longer senses that the aircraft is banking (i.e., he might mistakenly think the plane is moving forward on a linear course). Later, when the pilot exits the turn, stimulation within the semicircular canals can confuse him. He might mistakenly think that the plane is turning in the opposite direction rather than flying in a straight line. In response, the pilot might lean in the direction of the original turn in an attempt to compensate. Aviation accidents can and do occur because of such errors.

PHYSIOLOGICAL MECHANISMS OF HEARING

Historically, there have been several approaches to investigating functions of the inner ear apparatus. Selected lesions have been produced in experimental animals. Patients have been studied after a traumatic injury or surgical damage to the middle/inner ears. And unilateral versus bilateral surgical removal of selected components of the apparatus has been performed to determine its effects. Collectively, these approaches have yielded our current state of knowledge of the functions of the bony/membranous labyrinth in the inner ear.

In animals, the effects of the various techniques have been somewhat dependent on the species. For example, the rabbit seems to be one of the most sensitive species to such experimentation. When lesions are produced on one side of the head only, the rabbit turns its head, neck, and body in that direction. The unilateral lesion is associated with weakened musculoskeletal function on the affected side and strengthened function on the opposite side. This causes the animal to roll over and to turn near continuously in the direction of the lesion. Bilateral lesions appear to remove these effects as long as the eyes are open and proprioception is intact.

The cat is well known for its righting ability when free falling. In a state of free fall, the physiologically healthy cat will first right its head and neck. This is followed by righting of the forelimbs and torso. Finally, the hind limbs and tail are righted. If the height of the fall is sufficient, the cat will land on all four feet. Producing lesions on either side of the inner ear impairs the righting reflexes, causing the cat to fall on any part of its body. Moreover, if such an impaired animal is placed in a swimming pool, it fails to orient itself. Unless removed, it will drown.

In 1961, Georg von Bekesy was awarded the Nobel Prize in Physiology for his work on hearing. Among other accomplishments,
Bekesy developed a method for dissecting the inner ear of human
cadavers while leaving the cochlea partly intact. By photographing
strobe lights and markers in the tissue, he found that the basilar
membrane moved like a surface wave when stimulated by sound.
Different frequencies amplified the sound waves in different locations along the basilar membrane. High-pitch frequencies caused
more vibration at the base of the cochlea, while low-pitch frequencies created more movement near the apex. This is due in part to
the narrower, more rigid basal end of the basilar membrane compared to the apical end (which is wider and less rigid).

Bekesy also showed that the frequencies of sound waves are dispersed differentially before they excite various sensory nerve fibers
that lead from the cochlea to the brain. He theorized that the precise
location of each sensory hair cell along the coil of the cochlea cor-
responds to a specific frequency of sound. In addition, Bekesy developed a mechanical model of the cochlea that confirmed the concept
of frequency dispersion by the basilar membrane. Unfortunately, his
model could not provide any information about a possible function
of frequency dispersion in normal hearing or in hearing loss.

Many topics that are now current in hearing and sound are things
Bekesy could not have imagined. For example, the discovery of the
protein prestin, now known to be the main molecular motor for
movement of hair cells and cochlear amplification, were unknown
to Bekesy. The influence of the tectorial membrane, knowledge of
cochlear micromechanics, and the stimuli to inner hair cell stereocilia were all developed subsequent to Bekesy's experiments. Later, in
1985, other investigators showed that outer hair cells change length as
a function of changes in membrane polarization. Hyperpolarization
causes elongation of the outer hair cell, and depolarization causes

shortening. Moreover, the direct coupling of voltage and mechanics in outer hair cells exhibits a piezoelectric intelligence that is several orders of magnitude greater than the best-known piezoelectric materials manufactured by man in the 21st century.

Cochlear pathophysiology and the damage caused by aging, disease, noise, and traumatic injury were black boxes in experimentation on hearing before the 1980s. Where the field might go beyond him was as much a mystery to Bekesy as it was to the future of hearing research for those working in the early decades of the 21st century. Unfortunately, loss of hearing due to aging, disease, or injury is permanent because hair cells are irreplaceable. This is because aging, diseased, or accidentally damaged cochlea do not regenerate. Reduced numbers of progenitor cells and/or lower flexibility of the cellular epithelium resulting from the accumulation of calcified or glycosylated contractile proteins in cellular junctions are among the pathophysiological processes that might account for the loss of hearing. There are other explanations as well, including changes in blood flow and the coupling of electrical and mechanical activities of hair cells and the basilar/tectorial membranes.

BLOOD FLOW, ELECTROMECHANICAL
TRANSDUCTION, AND HEARING

As already mentioned, sound is a vibration or pressure wave traveling through a medium such as air, endolymph/perilymph, water, or blood. Sound pressure waves are converted into action potentials by the sensory organs and tissues of hearing. Action potentials are electrophysiological signals arising from changes in the polarity of excitable cell membranes, including the hair cells. When cell membranes depolarize, their polarity is lost. When cell membranes repolarize, the transmembrane polarity is regained. In the hearing apparatus, action potentials are conducted from hair cells to the brain via auditory neurons. Auditory neurons are connected to hair cells inside the cochlea.

In addition to membrane potentials and moving basilar/tectorial membranes, blood flow to the middle and inner ears plays an important role in hearing. The volume of cochlear blood flow is extremely small. In guinea pigs and rats, it has been estimated to be on the order of 1.5 µl/min (microliters per minute; it takes 1,000 microliters to equal one milliliter; it takes 1,000 milliliters to equal one liter or about a quart), or about one ten-thousandths of cardiac output. Also, cochlear blood flow is nonpulsatile and is anatomically distant from sensory hair cells (>100 µm), thereby minimizing perfusion-induced acoustic disturbances. In typical small rodents, blood flow velocity in various regions of the inner ear and adjacent tissues has ranged from 30 to 180 µm/s.

Normal blood supply to the cochlea is critically important for sustaining the production of endolymph. Abnormal cochlear microcirculation has long been considered a factor in noise-induced, age-related gradual loss of hearing (presbycusis) and in accidental acute loss of hearing or vestibular dysfunction. Endolymphatic hydrops is a condition that produces too much endolymph. It is characterized by fluctuating loss of hearing and periodic vertigo. Compared to healthy animals, reactive hyperemia is reduced by approximately a third in the ears of experimental animals with excess endolymph. Pressure-flow autoregulation of cochlear blood flow in guinea pigs with expanded volumes of endolymph is also reduced. Significantly higher circulating plasma concentrations of norepinephrine and vasopressin have been reported in patients with this condition. These and other chemicals (e.g., endothelin, angiotensin II, caffeine, nicotine) that constrict blood vessels might contribute to elevated tone of blood vessels in the inner ear, thereby reducing their autoregulatory capacity.

In student experiments at Rutgers University, we have observed that caffeine attenuates autoregulation of blood flow in the fingertips. Under these conditions, caffeine's vasoconstrictor properties are seen as early as 10–15 minutes after its consumption and reach marked levels by 45–60 minutes (*World Journal of Cardiovascular Diseases* 9:253–266, 2019).

At birth, the bundles of stereocilia on hair cells are immature. They lack the ability to convert mechanical stimuli into electrical signals. Mature, adult stereocilia are arranged in ordered rows that project from the surfaces of the inner and outer hair cell bodies. The structural and physiological properties of the inner hair cells change progressively along the cochlea. This design seems to ensure that each cell is tuned to selective frequencies of sound. For example, the bundles of hair cells are longest in the low-frequency cells at the apex of the cochlea and shortest in the high-frequency cells at the base.

Deflections of specific bundles of hair cells open mechanically sensitive ion channels in each cell membrane. Passing through these channels are depolarizing inward currents of charged ions. They change membrane potentials within the hair cells, and these changes cause synaptic vesicles to fuse with the hair cell membranes. After fusing, vesicles release neurotransmitter molecules. The neurotransmitter molecule crosses the synapse and binds to postsynaptic membrane receptors, causing changes in their membrane potentials. Action potentials are subsequently conducted through the auditory nerves to the brain for further processing of sound signals.

Since hair cells process a large volume of hearing-related information with accuracy and speed, the mechanical (bending of hair bundles) and electrical (action potentials) signals must be precisely integrated (electromechanical transduction). In this sense, stereociliary proteins are crucial for the development and maintenance of the structure and function of hair cells. Both the height and length of stereocilia are tightly regulated by the connecting and disconnecting of actin filaments. Actin is an important structural protein that is found throughout the body, including in all types of muscle. Mutations of prestin, actin, and other hearing-related proteins cause deafness in both mice and humans.

Signals from the auditory nerves are passed into the pons of the brainstem. In the pons, the calyx of Held of the auditory brainstem holds some of the electrophysiological mysteries to the central nervous system processing of sound signals. One of the primary functions of the calyx of Held is to localize the source of a sound. The timing of the activation of cochlear hair cells is important to this function and is especially critical in distinguishing low-frequency sounds in the horizontal plane. Because the structural and functional properties of the calyx of Held are markedly changed at the onset of hearing (e.g., postnatal days 10–15 in experimental rodents), most investigators choose to study neonates before this time. Marked, short-term depression of electrical excitability is regularly observed in every cell during in vivo recordings in tissue slices before postnatal days 10–15 (i.e., before the onset of hearing).

Three pathophysiological mechanisms seem to contribute to the short-term, prehearing depression of electrical excitability: (1) the depletion of presynaptic vesicles that have a high probability of being released (e.g., those closest to the synapse), (2) inactivation of presynaptic calcium ion channels, and (3) desensitization of postsynaptic membrane receptors. The relative contribution of each mechanism seems to depend on the stage of development and the pattern of stimulus. At low frequencies of stimulation, short-term depression is mainly presynaptic. Desensitization of postsynaptic membrane receptors is more important during prolonged high-frequency stimulation. At high rates of action potentials, delay at the calyx of Held synapse increases considerably. In the adult, few failures in synaptic transmission are seen in vivo.

Postsynaptic failures exist in the mature animal, but they are typically rare events. For example, they have been observed in the cat only during very high-frequency stimulation. In mice and rats, failures occur in approximately half of the cells but were observed only rarely in the gerbil. Hence there are considerable interspecies differences.

BALANCE, ORIENTATION, AND AGE-RELATED HEARING IMPAIRMENT

The physiology of balance, orientation, and vertigo is linked to both the apparatus and mechanisms of the inner ear. However, they are also influenced by altitude, oxygenation, proprioception, and vision. As mentioned earlier, in the laboratory, various experimental approaches can be taken to determine the specific roles of each tissue or organ on the maintenance of balance, orientation, and/or vertigo. These experiments have been done in both animals and in humans.

Some years ago, my family and I were waiting for a tour of the Mount Timpanogos Cave near Provo, Utah. The tour was late starting, our kids were hungry, and we had mistakenly left lunches in the van. My wife said, "Gary, quick, run down and get the lunches." I descended the distance in a few minutes (about 1,000 vertical feet; but following the winding trail filled with switchbacks, the distance to the parking lot was probably closer to a mile). I retrieved the lunches and then foolishly began jogging back up the trail. Midway, I was exhausted and got dizzy. I sat down to rest for a moment with my head on my knees. When I lifted my head, I could not distinguish up from down. Shortly thereafter, another tour group passed, and I asked them which way to the cave. They pointed the direction, so I continued, but more cautiously.

A couple months later, but before the first snowfall, my 17-year-old son and I decided to climb the nearly 12,000-foot Mount Timpanogos (we were in Provo, Utah, on sabbatical at Brigham Young University). We started at the trailhead about 5 a.m. in the dark. By 10 a.m., and now in the light, we reached the "Saddle" (about 10,000 feet). We stopped to rest. As I sat in awe of the Provo/Orem area nearly 6,000 feet below, I again became disoriented and dizzy. We still had 2,000 feet to climb in rougher, rockier terrain, and I was nervous that vertigo would overcome me. Still, we reached the summit about noon, ate our lunches, took some photos, and started the long hike back to the trailhead. As a lowlander acclimated to altitude and life at sea level, my vestibular apparatus was not accustomed to those altitudes or exercise at them. I probably experienced mild hypoxia, reduced delivery of oxygen to my brain, and the accompanying disorientation and vertigo that such aphysiological states can have on the inner ear and other sensory structures.

Trying to orient oneself after a thrilling ride at an amusement park is more difficult in the elderly than in the young. When getting off the ride, the elderly are more prone to stumbles and falls. If you don't believe this, consider visiting Six Flags Great Adventure in New Jersey. Take a ride on any of their giant, twisting roller coasters or Kingda Ka. At 456 feet, Kingda Ka is the world's tallest thrill-seekers ride. With a maximum speed of 128 miles per hour (in 3.8 seconds), it is also one of the world's fastest rides. For me, however, one of the most disturbing rides I have been on was Expedition Everest at Disney World in Orlando. After speeding forward for several seconds, the roller coaster we were riding inside a fabricated mountain came to an abrupt stop at the highest point on the tracks. It then reversed course and sped backward downhill and around curves at lightning speed and in the dark. This was disorienting and confusing for my vestibular apparatus. When the ride ended, my heart was racing, and I was imbalanced. I had to stand still supporting myself on a handrail for a time.

A good friend once asked me about my stretching and balance exercises. He then demonstrated an activity that reportedly improves balance. With eyes open, he took several steps, placing heel to toe in a straight line. Then with eyes closed, he tried to repeat the activity. Without the complementary use of vision and with impaired proprioception, his sense of balance was markedly upset. He could not take more than one or two steps without nearly falling over. I tried with a similar outcome. But as we have all been told, practice makes perfect or will usually improve one's ability. In my early morning exercises as I warm up, I rotate my head and neck 360 degrees repetitively, clockwise and then counterclockwise. I also walk several steps heel-to-toe. As long as I maintain this head-rotating activity several days per week, I hear and feel less crackling and popping of soft tissues and joints in my neck. Coupled with standing on one foot at a time while dressing/undressing, such activities seem to help maintain my sense of balance and orientation.

The next time you are in a group conversation, be attentive. Note the number of older members in the group (e.g., 60 years plus). Note also which age groups are doing most of the talking versus listening. If you are especially observant, you might notice older listeners leaning toward the speaker or in the general direction of the conversation. Because hearing is impaired with age, it is more difficult for older participants to hear the conversation. One who does not want to be excluded will do what it takes to be included (e.g., move his chair closer to the conversation, lean forward to move his head closer to the speaker). Thoughtful and courteous younger speakers/listeners/observers will make greater efforts to include the older participants.

As I try to age intelligently, I seek activities that will sustain my senses of hearing and balance. Instead of regularly turning the volume on my car radio up, I often turn it down and try to pay closer attention. When I am engaged in conversation, I also try harder to be a better listener. The fewer times I ask for clarification, the better I feel I am doing. I have also learned that as soon as the elderly begin falling, and especially if they begin breaking bones (legs, hips), death is not far behind.

4

PROPRIOCEPTION AND BALANCE

The word *proprioception* (or proprioception) comes from the Latin *proprius*, meaning "individual" (one's own, self) and *capio/capere*, "to grasp, grip, or take." Physiologically, proprioception is the sense of the relative position of one's body in space and time. It includes the sense of effort to move and the sense of acceleration. Proprioception can be distinguished from exteroception (perception of the world outside one's body) and interoception (one's sense of hunger, pain, etc., even though the sense of pain is more commonly referred to as nociception). Proprioception relies on sensors in muscles and joints. Proprioceptive information is essential for accurately guiding most movements, especially while they are being learned. Imagine the gymnast walking on the balance beam without the benefit of proprioceptors in the toes, ankles, knees, and other joints.

All voluntary movement is made possible by striated skeletal muscle. Skeletal muscles have two mechanically sensitive proprioceptors: muscle spindles and Golgi tendon organs. The rate of stretch of muscles and the actual length of a stretched muscle are monitored by muscle spindles. Golgi tendon organs measure the force generated by a muscle by determining how much stretch that muscle applies to the tendon to which it is attached. Working together, muscle spindles and Golgi tendon organs can produce a complete description of the dynamic work that a moving muscle is performing. The differences in sensitivity of the two sensors are determined not only by their unique anatomical structures but also by their placements in the body. Muscle spindles are located in modified muscle fibers called intrafusal muscle fibers. Intrafusal muscle fibers are arranged in parallel with the familiar force-generating muscles called extrafusal muscle fibers. Golgi tendon organs are arranged in series with extrafusal muscle fibers at or near the junctions of tendons and muscles.

Neurogenically, muscle spindles are connected to both afferent (sensory) and efferent (motor) nerves. Conversely, Golgi tendon organs consist of the endings of bare axons. These nerve endings interact with collagen and can be found at the junction of skeletal muscle and its corresponding tendon. When muscles develop tension by being either passively stretched or actively contracted, collagen fibers compress and distort nerve endings, causing them to generate action potentials. Muscle spindles do not, however, contribute significantly to the development of muscle force. They are strictly sensory organs. The muscle spindle contains two kinds of muscle fibers (bag and chain) and two kinds of sensory nerve endings (primary and secondary).

Besides the muscle spindle apparatus and the Golgi tendon organ, there are other kinds of proprioceptors. These are mostly present in the joints and can be found in ligaments and their connective tissue capsules. Some have well-defined morphology like that of the Golgi tendon organs, Pacinian corpuscles, and Ruffini end organs. Others are simply free nerve endings. All of them respond to changes in angles, directions, and velocities of movements in the joints.

SENSING LIMB POSITION AND MOVEMENT

The relationship between movement and position of one's limbs was originally described in the 16th century as a sense of locomotion. Much later, in the 19th century, the idea was expanded to include sensory activity within skeletal muscle. Muscle proprioception is now recognized as one of the first descriptions of physiological feedback. The basic notion was that commands, originating in the brain, were transmitted to the muscles and that reports on the muscles' condition were then sent back to the brain. In 1906, Charles Scott Sherrington published a landmark paper that introduced the terms *proprioception*, *interoception*, and *exteroception*. Beginning at that time and moving forward, physiologists and others have searched for specialized nerve endings that transmit mechanical information from joint capsules, tendons, and muscle spindles to the central nervous system. For their work in this field, Sherrington and Edgar Douglas Adrian jointly received the Nobel Prize in Physiology or Medicine in 1932.

From the above and subsequent discoveries, we now know that primary nerve endings in muscle spindles respond to both the magnitude and velocity of changes in the length of skeletal muscle fibers. We also know that muscle spindles contribute to the sense of both a limb's position and its movements. It has also been generally accepted that cutaneous receptors contribute directly to proprioception by providing accurate perceptual information about joint position and movement and that functionally, this knowledge is combined with information from the muscle spindles in producing responses.

In general, the ability to walk and to control one's posture are developed in early childhood and throughout adolescence. These activities reach their maximum levels of achievement in young adulthood and progressively decline thereafter with sedentary lifestyles and age. The musculoskeletal, neurogenic, somatosensory, vestibular, and visual organ systems critical to balance, posture, and walking are also developing and maturing during childhood and adolescence. They too decline with sedentary lifestyles and age. For example, it has been shown that if the visual and somatosensory systems are experimentally impaired so that the main sensory input is received from the vestibular system, both young children and older adults have difficulties with balance. Also, in experimental settings that require a rapid reevaluation of sensory inputs, both young children and older adults are at increased risk of losing balance.

Compared with healthy young adults, postural instability and incidents of falling are increased in both children and older adults. This seems to be due in large part to the attention individuals devote to maintaining their balance and posture while simultaneously paying attention to other concurrent activities (e.g., carrying a delicate or heavy object). The amount of attention needed to process postural information—therefore the less left over for a secondary activity, such as holding a cup of hot chocolate while walking—is an important indicator of one's ability to stand, walk, and fall.

In healthy older adults, the achievement of two or more simultaneously performed physical tasks can be improved by training. You can convince yourself of the truthfulness of this statement in your own home. When dressing/undressing, try remaining upright while standing on one foot. See if you can do this for 5–10 seconds (in a safe, cushioned space). Practice this over several days/weeks, and see if your time improves (e.g., to 10 or 15 seconds). Later, when your balance has improved, try standing on one foot and pulling on a sock without falling or losing your balance (dual tasks). Repeat this activity for several days/weeks until you have achieved it, and then switch to the opposite foot.

In New Jersey and like-minded states with little tolerance for drunk driving, law officers often require an impaired driver to attempt walking, heel-to-toe, in a straight line over a distance of several feet. Alternatively, they might ask the impaired driver to close her eyes and touch her nose. Both activities are compromised by alcohol, marijuana, and other harmful drugs. These compounds not only impede the central processing of somatosensory information; they also inhibit sensory receptors from performing their physiological functions. Alcohol and other drugs, sadly, are not the only things that impair drivers nowadays. More and more drivers of most ages below 60 (arguably) are increasingly impaired. These not-so-smart drivers are endangering all of us and encumbering the streamlined flow of traffic by trying to drive while tethered to their electronic devices—talking, texting, and reading.

Thus we humans go through our daily routines without being aware of the contributions of proprioceptors to our success and well-being. How inefficient would it be if, before every hand or arm movement, we had to prethink the next step? Most of us can put our index fingers to our noses, eyes closed, with great precision. But if we cannot see it, how do we know where our arm is in space and time as it moves from our side to our nose? And how does the index finger find the nose? To make things even more complicated, if the biceps muscle of the arm reaching for the nose is vibrated, sensations occur that suggest the arm and nose are lengthening. Is such a finger-to-nose maneuver more difficult to perform than a quarterback dropping a football into the outstretched arms of a wide receiver moving downfield at 8–10 meters per second and separated from him by 30–40 yards? Who has the greater proprioceptive task, the quarterback or the receiver? How confusing would it be if both players' biceps were vibrated while trying to throw or receive the football?

LOCATION OF PROPRIOCEPTORS: THE MUSCLE SPINDLE

Research on proprioception and proprioceptors has lagged behind work on the five basic senses because it is a sense we are largely unaware of. However, during the last 50 years, neuroscientists have taken advantage of new stimulation and imaging techniques to achieve further insight into this elusive yet essential sense, acquiring new knowledge both at the receptor level and on the central processing of proprioception.

Research on how we sense our body's movement dates back to Galen in the 2nd century AD. However, it wasn't until the early 19th century that specific ideas were introduced about the mechanisms underlying proprioception. German physiologists proposed that there was no need for peripheral sensory organs in proprioception. They believed that neurons in the brain drove muscle contractions. They also believed that these same neurons simultaneously sent copies of their signals to adjacent sensory areas of the brain to generate required sensations. The Germans called this theory a sensation of innervation. At the turn of the 20th century, English neurophysiologist Charles Sherrington challenged their idea on the grounds that we are aware of the positions of our limbs even when they are relaxed and immobile. Sherrington believed that there were sensory receptors in peripheral tissues that signaled position and movement. Today, elements of both theories contribute to the accepted view.

The most intelligent place to locate sensory organs that signal position and movement of a limb is in the muscles and joints of that limb. Similarly, the most sensible place to install shocks and struts on a car is near the wheels that hit the potholes, speed bumps, and other obstacles on the road surface. It would make no sense to place struts near headlamps or on door handles. Once they were located, and for many years after, it was believed that joint receptors were the principal proprioceptors. Recordings from central neurons during movements of a joint supported this notion. But there were other possibilities.

When the forearm is rotating about the elbow joint, not only is there movement at the joint (i.e., muscles inserting into the joint), but the elbow's flexors and extensors change length as well. Moreover, the length of muscles and other soft tissue of the upper arm are also influenced by movements of the forearm. In an important series of experiments published in 1972, Guy Goodwin and colleagues at the University of Oxford provided evidence that receptors in muscles, not in joints, were the most likely candidates for generating our sense of limb position and movement. Goodwin's lab showed that if the biceps muscle of a stationary arm of a blindfolded subject was vibrated, the subject perceived the arm as extending. The subject indicated the illusion by tracking the imagined movement with his other arm. Goodwin and coworkers argued that vibration had stimulated muscle spindles. The response to vibration appeared to simulate spindle activity, generated by the stretching of a muscle, leading to the illusion of a stretching biceps. The vibration of the triceps muscle led to sensations of the arm being flexed—that is, the illusion that the triceps was stretching. Importantly, vibrating the elbow joint failed to produce sensations of movement or displaced position. Hence the illusion could not be attributed to the responses of skin or joint receptors.

Muscle spindles are unique in that they have two kinds of sensory nerve endings. Primary nerve endings respond to both stretch and the rate of stretch of a muscle. Secondary nerve endings respond only to stretch—not to the rate of change of a muscle's length. Earlier animal experiments had revealed that primary nerve endings are especially sensitive to muscle vibration, while the secondary endings are vibration-insensitive. Also, in the early 1970s, investigators at the University of New South Wales showed that the illusion of arm extension was greatest with vibration frequencies of 80–100 cycles per second (cps or Hertz). When the frequency of stimulation was reduced below this range, the illusion of movement changed to one of displaced position. These investigators then proposed that there were two senses: the sense of limb movement, generated largely by primary nerve endings in muscle spindles, and the sense of limb position, generated by both primary and secondary nerve endings.

In 1986, French investigators showed that if the two antagonistic muscle groups acting at the elbow joint (the flexors and extensors) were vibrated simultaneously, the vibration illusion was eliminated. However, if the frequency of vibration of one antagonist was changed (i.e., either reduced or increased so that a difference in stimulation frequencies of the two muscles occurs) an illusion gradually began to reemerge, its magnitude being directly proportional to the difference in frequencies. These observations suggest that the brain does not process signals from each muscle in isolation but rather compares signals and computes arm position based on the difference.

Since the beginning of the second decade of the 21st century, other scientists have proposed that for this kind of arm-position matching task, it is not only the difference in signals from the antagonists of one arm that matters but also the difference in signals coming from both arms. When the difference is small, the arms are closely aligned. For example, Japanese investigators have shown that the size of the vibration illusion in one arm can be halved by vibrating the equivalent muscle of the other arm. Thus it seems that the brain is continuously monitoring movements of our arms relative to one another. Logically, the brain must do this to accurately align the arms and legs of, for example, gymnasts, who frequently use all four limbs in their difficult and precision routines.

Recently, investigators at Monash University in Australia tested the vibration illusion. However, rather than tracking the illusion with their free arm, subjects were asked to point at the perceived position of the vibrated arm that remained hidden from view. Surprisingly, subjects did not indicate any illusionary displacement of their arm during vibration. Yet when the more traditional matching task was used, the same subjects demonstrated normal vibration illusions. Furthermore, the characteristic position errors seen in a matching task, following a muscle contraction that had been attributed to muscle spindles, were no longer present in a pointing task. Considered collectively, these findings suggest that in a pointing task, muscle spindles no longer play the dominant role of position sensors that they do in matching tasks. That is, the source of the position signal changes depending on the nature of the task.

What sensory receptor, then, is involved in the position-pointing task? One possible candidate is a skin receptor, and rhythmic stretching of the skin enclosing a muscle can generate illusions of limb movement. However, when investigators tested this hypothesis, they found no evidence to support stretch receptors in the skin as the important sensor. As was hypothesized during the first half of the 20th century, sensory nerve endings in the joints are another candidate. There is evidence from movement-detection threshold experiments that joints, at least those in the fingers, can conduct sensory signals. The detailed role of joints, however, has not been studied in position-matching and pointing tasks, and this is a challenge for future studies.

Golgi Tendon Organs, Effort, and Force

In addition to knowing where our limbs are in space and time, at least two other sensations contribute to our physical self-awareness: a sense of force and a sense of effort or heaviness. When we are asked to compare the heaviness of two objects of nearly the same weight, we typically juggle them up and down in our hands before making a judgment. This suggests our sense of heaviness might be related to our sense of movement. When human muscles are treated with muscle relaxants, objects seem heavier, including the limbs themselves. This seems to result from an impairment in the sense of effort triggered by muscle weakness. In response, individual motor neurons fire at increased rates, and other motor units (a motor neuron and all the muscles it innervates) are recruited to generate increased muscle force. The combined recruitment and enhanced frequency of action potentials result in a greater perceived effort (i.e., the impression of increased weight). Similarly, in muscles weakened by exercise or overexertion, motor neurons increase their frequency of action potentials to compensate for the loss of force. That might explain why our limbs feel heavier at the end of vigorous exercise, especially in unconditioned individuals.

In a simplified view, the sense of effort is produced by impulses in the motor cortex that travel down the spinal cord to the lower motor neurons to trigger muscle contraction and relay back to sensory areas of the brain, where the sensation of effort is generated. However, there is evidence from experiments on fatigue, in which magnetic brain stimulation was used to mimic motor commands, that the sense of effort is generated somewhere upstream of the motor cortex and that the effort-force relationship undergoes constant adjustment. Furthermore, if neurons supplying a limb are blocked (e.g., from muscle relaxants or anesthetics) and an attempt to move the limb is made, sensations of changes in position and movement occur (in the absence of any). So the sense of muscle effort is linked to the senses of force, movement, and position.

LIGAMENTS, TENDONS, AND MUSCLE SPINDLES

Ligaments attach muscles to muscle, and tendons attach muscles to bone. At each end of a muscle is a tendon that anchors the muscle to bone. At the junction between muscles and tendons lie the Golgi tendon organs. Golgi tendon organs are small bundles of tendon strands enclosed in a connective tissue capsule similar to that of muscle spindles. Strands of tendon are composed of helical arrangements of interwoven collagen proteins. Each Golgi tendon organ has a large sensory axon that terminates on the tendon strands. The Golgi tendon organs respond to active muscle contraction, as well as passive stretch. The tendon organs send their sensory action potentials to the cerebral cortex with information about the amount of force the muscle has exerted, or the degree to which it has been stretched.

Golgi tendon organs are aligned in series with muscle and are exquisitely sensitive to the tension within a tendon. They are thus most sensitive to changes in muscle force and not to changes in the length of the muscle. Sensory axons from the Golgi tendon organ stimulate both excitatory and inhibitory interneurons in the spinal cord. In some instances, this interneuronal circuitry inhibits the muscle in which the Golgi tendon organ was activated and excites antagonistic muscles. The reflex activity of excited Golgi tendon organs is variable because the interneurons with which they interact also receive simultaneous inputs from nearby muscles and skin, as well as from higher cerebral cortical regions. Stability of joints and force within muscles are the kinds of reflexes that Golgi tendon organs mediate.

In experiments designed to illustrate the above principles, volunteers were asked to compare the stiffness of a series of compression springs. Subjects pressed the spring with one hand while using the other to select, from a range of springs, one with matching stiffness.

Under control conditions, volunteers were quite accurate in their choices. When muscles of the first hand were weakened by muscle relaxants, volunteers were still surprisingly accurate even though their compromised hands required more effort to compress the springs. When asked to match efforts and not forces, considerable errors were made in comparing stiffness of springs. It seems that we have the ability to choose between the senses of effort and force depending on the nature of the task.

Investigators have recently proposed that force signals of peripheral origin arise in both Golgi tendon organs and muscle spindles. If a muscle is progressively paralyzed, at the onset of the paralysis lifting an object is more difficult (it seems heavier). As the paralysis deepens, paradoxically, objects lifted seem to become lighter. This result has been attributed to the action of muscle spindles. When the muscle begins to weaken from paralysis, the spindles remain unparalyzed and their signals remain strong, contributing to the sense of increased heaviness. When paralysis progresses, the intrafusal fibers become paralyzed as well, leading to a reduction of the spindle signal. As a result, the object feels lighter than before. Thus in addition to a centrally generated sense of effort, we have a peripherally generated sense of force or of heaviness, which arises from signals in both muscle spindles and Golgi tendon organs.

In 1971, when he was 19, Ian Waterman suffered a bout of severe viral gastroenteritis. The illness triggered an autoimmune response that impaired his ability to orient his limbs in time and in space. As described by neurologists familiar with the case, Waterman was not paralyzed. His limbs could move, but he had no control over them. Proprioception, triggered by everyday activities, allows us to carry out tasks successfully without thinking about them. However, without feedback from proprioceptors, and like Ian Waterman, we would be confused, disoriented, and pretty much helpless.

Over the many months after he suffered the autoimmune-induced loss of proprioception, Waterman gradually learned to move again. At first, just standing stably was difficult. Using vision and a conscious will to move, Waterman, in his mid-60s at the time of this writing, is able to slowly combine muscle actions to achieve desired movements, such as lifting a cup of water. The more complicated the movement, the harder he has to think. Seeing his body and trunk are of critical importance. In the dark, he remains completely helpless. Remarkably, Waterman is also able to compare, as well as the rest of us, the heaviness of objects of identical size—provided his eyes are open. It seems that he judges the heaviness of objects by observing the speed and extent of their movement when he lifts them. Waterman's unique case emphasizes the importance of proprioception in our daily lives.

PROPRIOCEPTION AND EXTREME SPORTS

Our sixth sense not only allows us to control the movements we make; it provides us with a sense of self, the awareness of our body's acceleration and position as we navigate surroundings. Nonetheless, proprioception can be challenged to the limits, and extreme sports illustrate this. The Flying Wallendas Family has been pushing proprioception to its limits for generations. Nik Wallenda, a seventh-generation performer, earned his reputation performing daredevil high-wire walks over Niagara Falls, a gorge near the Grand Canyon, between Chicago's skyscrapers, and across the rim of the Orlando Eye observation wheel. On one occasion, five of his performing associates were rushed to area hospitals after tumbling 25–30 feet from a high wire. The performers were attempting an eight-person pyramid during a practice session when they lost balance and fell to the ground. Six men and two women were part of the act, and

Wallenda said he was among the three who were able to catch the wire and avoid falling.

In 1962, two members of the Flying Wallendas died, and another was paralyzed after a performer lost his balance during an attempt at a seven-person pyramid in Detroit. Later, in 1978, Karl Wallenda died on a high-wire walk between two towers in San Juan, Puerto Rico. Other stunts that defy proprioception have left daredevils maimed, paralyzed, or dead. Family legacy or not, pushing proprioception to the limits that the Wallendas and others do has always been and will continue to be dangerous and, some would argue, unintelligent.

In 1995, ESPN introduced the X-Games, and two years later, they founded the Winter X-Games. These activities soon became the launching pads for extreme sports such as snowboarding, skateboarding, wakeboarding, and surfing. Extreme bike sports like BMX, Downhill Mountain Biking, and Freestyle Motocross were soon shot to the forefront of all action sports. By the time of this writing, circa 2019, such activities have taken over the ranks of all recreational sports and have become mainstream. More youth, young adults, and weekend thrill seekers are taking part in extreme activities each year. The attraction lies in the individualism of the activities, an increasing love for doing sport in the great outdoors, and the endless drive of challenging oneself to the extremes of performance.

To more fully engage in the activities, and realizing the mental, physical, and physiological levels of difficulty, many participants are turning to exercise scientists and coaches for help with development and training. Such training requires more than mental stamina and determination. It requires extremes in repetition as well, and there is a corresponding increased risk of injury when extreme repetition leads to exhaustion and weakness. Exercise professionals face the challenge of preparing competitors to push themselves to physical limits that are often beyond reason, dangerous, and even life-threatening. Trainers must be knowledgeable of biomechanics, the analysis of time and motion, and basic physiology in order to create new moves and techniques.

Snowboarding extremes require the athlete to catch air (go airborne), grab the board, perform 180° to 1,080° spins (half to triple spins), and switch stances—all while moving in a forward direction and bending at difficult angles. Freestyle snowboarding requires participants to navigate half pipes, super pipes, and boardercross maneuvers. These include descending/ascending banks of snow of 15–20 feet in height—separated by a distance of only 15–25 feet (the famed U-shape)—and going airborne all while performing a variety of tricks.

Motocross might even require more skill and could be more dangerous. For one thing, competitors ride motorized vehicles burning flammable fuels in powerful engines that jettison athletes 10–40 feet into the air. As their bikes are launched, riders perform airborne spins. This reduces body contact with the bike (e.g., taking one or two feet off pedals), often separating rider and bike at altitudes of 30 or 40 feet. Bicycle Motocross (BMX) originated with teenagers trying to imitate their older Motocross idols on nonmotorized bicycles. It involves similar risks and challenges using the same muscles, bones, and proprioception.

The unstable and unpredictable environments of extreme sports demand extraordinary skills of balance and strength. Personal agility, coordination, dynamic balance, and proprioception are all critical factors to success in extreme sports. For example, better balance leads to improved control. Improved control enhances visual awareness, which results in better decision-making. All this translates to success on the course and increased likelihood of positive outcomes such as rider safety, first-place standings, and continuing competition.

Proprioception and Development across Species

The benefits of touch have become a recurring theme in human, monkey, and rat development, but the importance of physical contact is not limited to mammals. Microscopic nematodes, known to scientists by the abbreviated name, C. elegans, are also sensitive to touch and touch deprivation. Nearly half a century ago, Sydney Brenner chose C. elegans as a model species for experiments in developmental biology and neuroscience. Now the worm is one of the best-understood invertebrate organisms on earth. Morphologically, C. elegans is more than simple. The adult worm has fewer than 1,000 cells, about 300 of which are neurons. These neurons form approximately 5,000 chemical synapses, 2,000 neuromuscular junctions, and 600 gap junctions. Its small size (approximately one millimeter) and short life cycle (fewer than three days) have assisted investigators in creating a complete neural wiring diagram and cell lineage in this species. Moreover, the ability of C. elegans to survive freezing temperatures for long-term storage makes it highly amenable to laboratory research.

The worm develops from an egg to an adult in three days, passing through four larval stages (L1–L4) before entering young adulthood. It has been shown that C. elegans reared in isolation have a smaller body size and a delayed onset of laying eggs when compared with worms reared in colonies of 30 or more. Growing up alone on a smooth agar surface in an insulated incubator, isolated nematodes, similar to premature C. elegans, were deprived of mechanosensory stimulation. As a result, they showed significantly smaller withdrawal responses (backward swimming) to a tap delivered to the side of their petri dishes. Investigators have also found that the reduced body size of isolated worms could be partially reversed if they were transferred into colonies before the end of stage L3. This demonstrates that exposure to other worms alters adult body size and that there is a critical period (stages L1 to L3) for such exposure to be effective.

In 1922, in one of the earliest studies on the benefits of touch, investigators reported that infrequently handled rats were more timid, apprehensive, and hyperactive than rats that had been handled regularly. Such pups were also six times less likely to survive thyroidectomy. For the developing rat pup, dams and littermates are the major sources of sensory input. Thus one useful approach for evaluating the importance of mechanosensory input is to remove it completely and observe what happens in experimental animals.

In keeping with this hypothesis, investigators have compared the adult behavior of maternally reared rats with that of pups reared in isolation from postnatal days 4 to 20. Despite receiving comparable nutrition, pups raised in isolation weighed less at weaning. Although this difference did not persist into adulthood, early deprivation affected adult maternal and emotional behavior. For example, as adults, those reared in isolation were less attentive to their own offspring. They engaged in fewer retrievals of their pups and spent less time licking and crouching over them. Conversely, more time was spent digging, rearranging bedding, and biting the cage and hanging from its sides and tops. They also spent more time chasing their tails and seemed more concerned about themselves than their offspring.

In an attempt to reverse the effects of isolation on adult behavior, pups reared in isolation have been stroked with a warm wet paintbrush to simulate maternal licking. Minimally stimulated pups received 45 seconds of anogenital stroking twice a day to promote urination and defecation, while maximally stimulated pups received two minutes of full-body stroking five times per day. When these pups were studied as adults, and when the way they mothered their own offspring was examined, it was found that full-body stroking partially recovered the behavioral deficits of isolation. Maximally stimulated pups exhibited maternal behavior of durations intermediate to those of maternally reared and minimally stimulated pups. Thus tactile stimulation can ameliorate some of the deficits resulting from isolation, but nothing can replace the time and contact a mother devotes to rearing her offspring.

Twenty preterm human neonates were transferred from a neonatal intensive care unit to a transitional care unit and were then provided 15 minutes of mechanosensory stimulation three times per day for 10 days. The procedure was tightly controlled, as infants received body stroking for the first and final 5 minutes of stimulation. During the intervening 5 minutes, their limbs were gently flexed upward. Several clinical and behavioral variables were monitored, and the stimulated infants were compared with unstimulated controls equivalent in gestational age (approximately 31 weeks), birth weight (approximately 2.8 pounds), and duration of neonatal intensive care (approximately 20 days). The extra mechanosensory stimulation led to superior growth and developmental performance.

Although diet and nutrition did not differ between the two groups, infants receiving additional human contact averaged 47 percent greater weight gain per day than those denied the extra contact and were discharged an average of six days earlier. The stimulated infants also spent more time awake and active and exhibited more mature habituation, orientation, motor, and range-of-state behaviors. Moreover, the positive effects appeared to persist. When retested 8 and 12 months after treatment, the stimulated infants were in a higher weight percentile group, scored better on mental and motor assessment tests, and had reduced incidences of minor neurological abnormalities.

SKIN, HUGS, AND PROPRIOCEPTION

Skin is the body's protective covering and our largest sensory organ. Unique among our sensory systems, sensory receptors in the skin give rise to distinct sensations, including affection and gentle touch, pain (nociceptors), itch (mechanoceptors), and warmth and cold (thermoreceptors). These distinct sensations are initiated by an impressive array of somatosensory afferent neurons that densely innervate the skin. Humans and other animals rely on sensory inputs from the skin to interact with others and with objects in their environments (including avoiding harmful interactions). Our sense of touch allows us to perform numerous activities that rely on fine motor control. Examples include word processing (typing), feeding, and dressing. Touch is also important for social exchange, including bonding with others.

As illustrated above, neonates deprived of touch (e.g., holding, cuddling, hugging of human infants) display developmental and cognitive deficits. Several landmark studies have highlighted the extremes of such deprivation. Also as mentioned above, the impairments can be improved and at least partially overcome by daily contact with others. However, cognitive deficits in rodent pups deprived of touch persist through adulthood, highlighting the importance of touch during early development.

In today's world of infant harnesses and slings, safety seats and restraints, and bus-sized, live-in strollers, it is a wonder that infants get touched at all except for the obligatory transfers from one space to another. Such devices and equipment have been hastily developed for the pretense of safety and protection with little or no regard to the fundamental importance of cuddling, holding, rocking, and otherwise touching. Generations-old parental practices such as carrying and holding infants in one's arms are endangered activities.

The endangerment is enhanced by the additive complications of dual-family incomes, ubiquitous daycare centers, and preoccupied parents pursuing careers, additional degrees, income, and personal gratification.

A Family Man is a 2016 American drama film directed by Mark Williams and written by Bill Dubuque. The film stars Gerard Butler as Dane Jensen, a driven—at times immoral—Chicago-based corporate headhunter disengaged from his wife and their three children. Dane's life revolves around closing deals, working late hours, missing important family events, and generally being a jerk on the home front. Willem Dafoe (Ed Blackridge, Dane's boss) owns the Blackridge Recruitment agency. Blackridge was never married, has no children, and is motivated only by boatloads of money and driving employees to extremes.

Dane's oldest child, Ryan (Maxwell Jenkins), gets leukemia, is hospitalized, and nearly dies. This partially awakens the disengaged father and husband but only periodically. Ryan is about 12 years old and has a younger sister about 6 and brother about 2. Only on one or two occasions during the nearly two-hour film does Jensen interact with his 6-year-old daughter. He is completely unaware of his 2-year-old son (the film shows no interactions of Jensen with him). The children's mother, Elise Jensen (Gretchen Mol), does her best to keep the family together during these difficult times. She could use daily relief from her husband but never gets it.

The film, I think, accurately portrays many of today's absentee fathers. Between seeking more education and advanced degrees in their 30s, 40s, and 50s, and by devoting inordinate time and attention to careers, income, and the stock market, such fathers are almost completely disconnected from their wives and children. They are fathers in absentia. Dane Jensen of *A Family Man* is only one good example of this misfortune.

I knew a husband and wife who grew up in quite different families. The husband's mother was a hugger and a kisser. She lavished these acts of affection and love on her children at all ages, infancy to parenthood. The wife's mother was quite the opposite. At least in private settings and family gatherings I witnessed, she rarely touched her children (also from infancy to adulthood). She exemplified the absence of proprioception described above. Hugs were not experienced by her children, and as they matured, being hugged and giving hugs were difficult for them.

The reason the hug is important, and at least one intelligent use of hug-related proprioception, is to send and receive unmistakable messages of caring and love. Unfortunately, however, in today's world there are at least two kinds of hugs. The traditional hug, characterized by full-body contact, sends the messages "I love you," "I have missed you," "Welcome home," "You are valued," and so on. It is an endangered species. Anything other than this is a pseudoembrace, an "A-frame encounter." This is where, upon meeting/greeting, two people lean passively and almost apologetically against each other (in stiffened-body, A-frame stances). Heads are ducked and turned awkwardly as corresponding shoulders accidentally meet. Arms dangle flaccidly at one's side or on the other's shoulder, but at all costs, chests and torsos do not come in contact. Proprioceptors in backs, joints, muscles, and skin are less engaged in this stance, as we send unintelligent messages of discomfort and uneasiness. Such awkward and unnatural yet politically correct interactions say, "I really don't know what to do, or if I even care about you."

Sadly, as the world grows ever more dangerous, we become increasingly suspicious of each other. We are decreasingly emotionally connected. The will and the ability to give and receive messages of love and affection are being lost. We imagine pedophiles, sex offenders, and personal-identity thieves skulking everywhere, and we favor political correctness over emotional and intellectual connectivity. Such imaginations are wrong, and they are dangerous. They explain, in part, why we purposefully avoid eye-to-eye contact and pretend to be talking on our smartphones as we pass each other on the sidewalks. Proprioception beware!

5

ACID, BASE, AND
pH REGULATION

The activity of hydrogen ions in a solution (where a solution is a combination of solutes dissolved in a solvent) determines its acidity. Activity of hydrogen ions is expressed in the dimensionless unit: activity of hydrogen ions is a_{H+}. The a_{H+} in pure water is about 10^{-7}. By convention, solutions having an a_{H+} greater than 10^{-7} are acidic solutions, and those having a_{H+} less than 10^{-7} are alkaline, or basic, solutions. In the human body, a_{H+} ranges from about 0.13 in the most acidic gastric juices to about 0.00000003 in the most alkaline pancreatic fluids.

In order to find a reasonable expression for this wide range of activity of hydrogen ions, acidity is expressed in terms of the negative logarithm to the base 10 of the activity of hydrogen ions in a solution. The term used is *pH*, and its definition is expressed by the equation $pH = -\log a_{H+}$. In general chemistry courses at all levels, students learn about the pH scale. This scale of acids and bases ranges from 0 to 14, with 7.0 being neutral. *Neutral* means that the contributions of acids and bases to the pH of a solution are equal.

Another purpose of this chapter is to help the reader understand an intelligent outline of the ABCs of physiological acid-base chemistry. The topic is of such tremendous practical importance that its serious study is near-mandatory. Sadly, for some there are no effortless roads leading to knowledge of acid-base chemistry, and there are no easily memorized rules that can be applied generally without risk of catastrophic outcomes. Consequently, students or readers wishing to master the subject must work earnestly and seriously. Once one has mastered the basic principles, then the more difficult aspects one might encounter in life, in research, or at a patient's bedside are more easily grasped.

Our body fluids are solutions—suspensions of water (the solvent) and organic and inorganic chemicals (the solutes). Organic chemicals, also called *hydrocarbons*, have a backbone structure made of carbon and hydrogen atoms with other elements attached to these. All other chemicals are inorganic. A solution is nothing more than a solvent with solutes in it, and water is the universal solvent—that is, more solutes dissolve in water than in any other solvent.

In humans, body water is found in three well-defined spaces (also called *compartments*). They are (1) the intracellular space, (2) the extracellular space, and (3) the transcellular space. Examples of intracellular water can be found in secretory cells of salivary glands, muscle cells, nerve cells, and many, many other cell types. Extracellular water resides in two subspaces: the vascular space (also called the *vascular lumen* or the *intravascular compartment*) and the interstitial compartment. Blood is usually confined to the intravascular space and is by weight and volume approximately 90 percent water.

The other extracellular compartment is the interstitial space. The prefix *inter-* means "between," and the suffix *-stitial* means "tissues." Thus interstitial space is found surrounding and outside of cells and tissues. When I think of the body's interstitial spaces and the processes that take place there (most of them still a mystery), words like *interstellar* and *interlithic* spaces come to mind. These are the mysterious spaces between the stars and between the rocks and soils of the earth. All three spaces, interstitial, interstellar, and interlithic, hold secrets that science is clueless about, mostly because such spaces are inaccessible and therefore immeasurable. This does not mean, however, that important events affecting our world and lives are not taking place there.

Transcellular fluid is found in a much smaller compartment called the *transcellular space*. This space characterizes bone joints (e.g., of the knee and elbow) and the synovial fluid contained therein; cerebral ventricles and the spinal canal (which houses cerebrospinal fluid); anterior and posterior chambers of the eye (aqueous and vitreous humors, respectively); the pericardium (a membranous sac that contains the heart and pericardial fluid); intrapleural spaces (membranous sacs that contain each lung and the intrapleural fluid); and the peritoneum (the abdominal cavity and peritoneal fluid). One characteristic that makes the transcellular compartment different from intracellular, intravascular, and interstitial spaces is that the fluids it contains are completely enclosed by two layers of epithelial cells. Like the interstitial, interstellar, and interlithic spaces, transcellular spaces are technically difficult to instrument and experiment on. Once excess water accumulates in the transcellular spaces, it can be difficult or impossible to remove.

The pH of the fluids in all these spaces must be kept in physiological balance with adjacent spaces. For example, the pH of arterial blood in a healthy human is about 7.40, while that of venous blood in the same person is about 7.36. (That is, venous blood is normally slightly acidic compared to arterial blood.) Only modest and temporary changes in pH can occur in either vascular compartment in order for a person to stay in good health. If the pH of one's arterial blood rises above 7.40 (e.g., 7.60 or higher), that person is said to be *alkalotic* (having blood more basic than normal). If it declines below 7.40 (e.g., 7.20 or lower), he or she is in an *acidotic* state. Neither of these pH conditions is consistent with sustained good health, and departures of only a few hundredths of a pH unit from the normal range can cause death after a few hours.

BUFFERS AND BODY WATER

Another definition of an acid is any compound capable of giving up or donating a hydrogen ion (or a proton). A *base* is any compound able to take up or to receive a hydrogen ion or a proton. In our bodies, one common acidic group is the carboxyl group (R-COOH). The letter *R* represents any other chemical group that might be attached to the carbon atom (*C*). When this compound dissolves in body water, it gives up the –OH group to yield a hydrogen ion and an anion (negatively charged group). Since this anionic group is capable of accepting another hydrogen ion, or positive cation, the anion becomes a base. In this case, such an anion is known as the *conjugate base* of the corresponding acid. Because the acid and its corresponding conjugate base are able to minimize changes in the hydrogen ion concentration of a solution, the pair act together as a buffer.

In the body's aqueous compartments, a buffer is a chemical or physiological event that corrects or adjusts changes in pH. Chemical buffers are atoms and molecules found in the extracellular, intracellular, transcellular, and renal tubular spaces. Physiological events intended to buffer body fluids occur in the kidneys, lungs, and other organs such as the pancreas gland. Thus when the need arises, the kidneys can excrete urine that is more or less acidic or basic, and the lungs can exhaust more or less carbon dioxide (CO_2), a gas that can alter the pH of body fluids and change acid/base homeostasis. The pancreas produces bicarbonate ions to buffer excess acids released from the stomach into the small intestine.

The actions of a buffer can be described quantitatively and graphically in the form of a titration curve. Such titration curves are constructed by dissolving known amounts of buffer in water and subsequently measuring the change in pH of the buffer/water solution. Then a known amount of acid is added or removed, and the solution's pH is again measured. The process is repeated until the entire buffering range of the buffer is covered. To construct the relationship as a curve, the amount of acid that has been added or removed is plotted on the vertical axis (the ordinate) as a function of the pH plotted on the horizontal axis (the abscissa). Such a curve has a negative slope that decreases from left to right along the x-axis. The buffering value of a solution is the quantity of hydrogen ions that can be added to, or removed from, a solution with a corresponding change of 1.0 pH unit. This value is determined by the slope of the titration curve as described above.

It is difficult for this writer to continue without identifying a few chemicals by their atomic and molecular formulas (not their molecular structures, a related but different concept). Understanding these chemicals and pronouncing their names need not be intimidating even to the nonchemist or uneducated. Having said that, the most effective extracellular buffer is the carbonic acid / bicarbonate ion (H_2CO_3/HCO_3^-) system. The carbonic acid / bicarbonate ion buffer system derives from the basic concept of molecular interactions between water and carbon dioxide. Carbon dioxide is just one of an infinite number of by-products produced by working cells, and water is ubiquitous throughout the body.

When carbon dioxide and water come in contact, the gas molecule (CO_2) gets hydrated by the water molecule (H_2O), producing the weak carbonic acid (H_2CO_3). *Weak* means the acid dissociates slowly to its composite parts: hydrogen ions and carbon dioxide gas. A *strong* acid is one that dissociates rapidly, thus producing many hydrogen ions in a short period of time (e.g., hydrochloric acid, HCl, is a strong acid). As long as carbon dioxide is produced and hydrated by water, H_2CO_3 is synthesized and subsequently dissociates to yield free hydrogen ions (H^+) and free bicarbonate ions (HCO_3^-). Adding hydrogen ions to an aqueous solution makes it acidic, while adding bicarbonate ions makes it basic. Thus adding free hydrogen ions will decrease the pH of a solution, and adding bicarbonate ions will increase the pH.

The H_2CO_3/HCO_3^- system is not an intracellular buffering mechanism. The best intracellular buffer consists of proteins and the dihydrogen phosphate / hydrogen phosphate ($H_2PO_4^-/HPO_4^{-2}$) system. This system (except for proteins) is also effective in renal tubular fluid, where it becomes concentrated because more water than dihydrogen phosphate / hydrogen phosphate is reabsorbed from the tubules. The reason this system is effective in renal tubular fluid is because filtered bicarbonate is generally reabsorbed by the renal tubules, leaving an excess of unbuffered hydrogen ions. The dihydrogen phosphate / hydrogen phosphate system is the first line of buffering defense against this excess of filtered hydrogen ions. Whenever excess hydrogen ions in renal tubular fluid are buffered by a system other than the carbonic acid / bicarbonate ion system, it is the equivalent of adding an additional bicarbonate buffer to the blood.

If either a strong acid or a strong base is added to renal tubular fluid, the dihydrogen phosphate / hydrogen phosphate system reduces the strength of the acid or base by converting it to a weaker acid or base. This action yields only modest, as opposed to marked, changes in pH in the tubular fluid and averts potential pH disasters. Because of its physical chemistry, the dihydrogen phosphate / hydrogen phosphate system operates at near-maximal capacity at a pH of 7.40, or near neutral for most body fluids. On the other hand, its concentration in the extracellular spaces is so low that it operates there at only a few percent of the carbonic acid / bicarbonate system. Thus when compared to carbonic acid and bicarbonate ions, dihydrogen phosphate and hydrogen phosphate are of minimal value in buffering acids and bases in the interstitial spaces.

Other important intracellular buffers are proteins. Proteins are highly concentrated inside cells and are one of the most abundant buffer systems in the body. Even though the pH inside the cell is modestly less than that outside the cell, intracellular pH changes in proportion to fluctuations in the pH of extracellular fluids. The rapid diffusion of carbon dioxide and the much slower diffusion of hydrogen and bicarbonate ions across cell membranes mean that changes in pH inside or outside the cell can affect one another. Changes in pH caused by diffusion of carbon dioxide are rapid, while those caused by hydrogen and bicarbonate ions take much longer (except inside red blood cells). Proteins, including hemoglobin, combine with carbon dioxide and hydrogen ions. When combined with carbon dioxide, the resulting protein molecules are called *carbamino compounds.*

Hemoglobin is the most concentrated protein inside red blood cells. Whenever oxygenated hemoglobin (HbO_2) releases oxygen to active cells, it leaves deoxygenated hemoglobin behind. Active cells simultaneously produce CO_2, which diffuses into the same blood supply that just gave up oxygen. The release of oxygen frees up a hemoglobin molecule to buffer either CO_2 or H^+ (simultaneously produced by the hydration of CO_2 and resulting in H_2CO_3, which dissociates to yield HCO_3^- and H^+). Hemoglobin's transport of oxygen to the tissues and carbon dioxide away from them is an act of intelligence. Lack of oxygen and overabundance of carbon dioxide, which would occur in the absence of hemoglobin, are potential problems. In the presence of hemoglobin, the problems are solved.

More specifically, because hemoglobin inside the red blood cell of the arterial circulation is on the alkaline side of the pH scale, its number of negative charges is greater than its number of positive charges, giving the hemoglobin of arterial blood a net negative charge. Addition of hydrogen ions to the buffer groups of hemoglobin (most notably the imidazole group) increases the number of positive charges and decreases the number of negative charges. Thus hemoglobin's net negative charge is reduced as it arrives at the peripheral tissues, where actively respiring cells produce and release hydrogen ions that need to be buffered. These changes in hemoglobin's structure and charge are what give it the capacity to unload oxygen, to upload acid, and to act as an effective buffer in the tissues.

An analogy likening the intelligence of unloading oxygen and uploading carbon dioxide in the human body by hemoglobin to a well-known business model might be instructive. It makes sense to carry a load from point A to point B, then another from point B to A, rather than to return empty.

In transporting and exchanging goods and services between two points, whether from inside to outside a cell or between two or more organ systems, our bodies would never make such a wasteful mistake (unless, of course, they are impaired by injury or disease). The hemoglobin molecule never goes home with an empty load. Rather, it returns with carbon dioxide and hydrogen ions ("merchandise") to the lungs ("New York City"), where it will unload gaseous byproducts before picking up a life-sustaining new load, binding fresh oxygen from the alveoli and returning it to the respiring tissues ("San Francisco") (see also *Our Marvelous Bodies*, Rutgers University Press, 2008).

Hemoglobin, bicarbonate ions, and the dihydrogen phosphate / hydrogen phosphate system are all chemical buffers. As molecules they act as buffers by combining with other ions (e.g., hydrogen) or molecules. The next sections focus on nonchemical buffer systems. The kidneys and lungs are organs that respond to changes in the pH of body fluids. They too help protect the homeostasis of acid/base balance by buffering acids and bases. Even though standard physiology textbooks describe these organs as nonchemical buffer systems, they do indeed operate by changing the chemistry of body fluids.

KIDNEYS AND ACID-BASE BALANCE

The kidneys perform four vital functions: (1) they filter blood plasma, (2) they reabsorb and secrete chemical products, (3) they excrete waste products in the urine, and (4) they can perform multiple other functions. For example, kidneys help regulate the production of red blood cells in bone marrow. They also assist in the control of blood pressure. The kidneys help regulate pH of body fluids by excreting acids and bases in the urine (involving items 1–3). Excreting an acidic or basic urine reduces the concentrations of these substances in the extracellular fluids (including blood), thereby helping to maintain acid/base homeostasis.

The actual quantity of a solute filtered by the kidneys each day is a product of the concentration of a solute in the arterial plasma flowing to the kidneys and the rate of filtration taking place in the glomerular capillaries. This product is called the *filtered load* of the solute. For example, if the glomerular filtration rate (GFR) is 180 liters per day (L/day) and the concentration of bicarbonate ion in renal arterial plasma is 24 milliequivalents per liter (mEq/L), then the product of these two numbers is 4,320 mEq/day, or the actual number of bicarbonate ions filtered by the kidneys in a day. In order for this filtered load of bicarbonate to be reabsorbed and not wasted in the urine, the kidneys must secrete an equivalent load of hydrogen ions downstream to the site of bicarbonate filtration. Thus from the peritubular capillaries that lie downstream to the glomerular capillaries, hydrogen ions are secreted at a rate of 4,320 mEq/day into the renal tubules.

If the kidneys excrete an excess of acidic urine, then the fluids in the extracellular spaces will be basic. If they excrete basic urine, the extracellular spaces will be acidic. The overall mechanism by which the kidneys help maintain acid/base homeostasis is by employing the carbonic acid system described earlier. When equivalent loads of filtered bicarbonate and secreted hydrogen ions meet inside the renal tubules, they combine to produce an equivalent mass of carbonic acid. This is called *renal tubular titration* of acids (hydrogen ions) and bases (bicarbonate ions). The carbonic acid then dissociates to yield water (for excretion in the urine) and carbon dioxide (for reabsorption by renal tubular cells, where it is then rehydrated and reabsorbed into the renal interstitium and blood).

Because there is an excess of hydrogen ions compared to bicarbonate ions in tubular fluid, renal tubular titration is not exact. The excess hydrogen ions (about 80 mEq/day) come from nonvolatile acids produced by the continuous degradation of proteins both dietary and structural/functional (from our cells). For example, muscle metabolism continuously releases the protein creatine into the blood. Degradation of creatine and other by-products of muscle metabolism produces nonvolative acids A volatile acid can be converted to carbon dioxide and then exhaled by the lungs. A nonvolatile acid is not converted to carbon dioxide and must be excreted by the kidneys.

Metabolism of creatine yields another by-product: creatinine. As long as muscle mass does not change significantly over extended periods of time, the rates of production and circulating concentrations of creatinine remain relatively constant. Because of this, both the ability of the kidneys to clear creatinine from the blood as well as the actual circulating concentrations of creatinine in the plasma can be used as indicators of kidney function and health. If you look closely at the laboratory reports from your most recent physical examinations (over the last 10 years or so), you will find in the blood chemistry / blood workup the acronym *GFRe*. This means *estimated glomerular filtration rate*, and it is based on the circulating plasma concentrations of creatinine. Review your recent lab reports or watch for your GFRe in the next report. This is just one example of the ongoing discovery by physiologists and clinicians that the body produces markers that can be used intelligently to evaluate the status of an organ's health. I will mention others as we continue.

Another chemical buffer system in tubular fluid that the kidney uses to maintain pH is the ammonia / ammonium ion (NH_3/NH_4^+) system. This system is even more important than the dihydrogen phosphate / hydrogen phosphate system mentioned earlier. Ammonium ions come primarily from the metabolism of amino acids, such as glutamine, in the liver. Renal tubular sites for this system are found in the proximal tubules, the ascending limbs of the loop of Henle, and the distal tubules. To comprehend the renal tubular system, and for further morphological detail, one can consult a good book of anatomy to review the structure of nephrons (the functional unit of the kidney). For example, the *Atlas of Human Anatomy* by F. H. Netter is a good place to begin (sixth edition, Elsevier).

Once inside the epithelial cells of the tubular nephron, glutamine is reduced to ammonium and bicarbonate ions through a series of metabolic reactions. For each molecule of glutamine metabolized, two ammonium ions are secreted into the urine and two bicarbonate ions are reabsorbed into the interstitium and blood of the peritubular capillaries. This system is therefore a continuous generator of new bicarbonate ions. In the collecting tubules, hydrogen ions are secreted into the lumen, where they combine with ammonia (which easily diffuses through the tubular epithelium) to produce ammonium ions. Ammonium ions are nondiffusible and therefore get trapped in the lumen and excreted in the urine.

Lungs and Acid-Base Balance

The lungs are also vital organs for regulating pH of body fluids. This is done through the operation of respiratory chemoreceptors. A *chemoreceptor* is a sensory organ or tissue that can respond to changes in its chemical environment. There are two populations of respiratory chemoreceptors that help regulate the rate and depth of ventilation. The rates and depths of ventilation in turn influence the acid/base status of the tissues and body. One population of respiratory chemoreceptors is found in the walls of major blood vessels (called *peripheral chemoreceptors*), and the other is found outside the cerebral blood vessels in the interstitial spaces (called *central chemoreceptors*). Both sets of respiratory chemoreceptors are capable of responding to changes in their chemical environments. The main chemicals that are detected by these sensory receptors are the partial pressure of oxygen (PO_2), the partial pressure of carbon dioxide (PCO_2), and pH (or the hydrogen ion concentrations of the body fluids bathing the receptors).

The peripheral respiratory chemoreceptors are located in the carotid sinus and the aortic arch and are called *carotid bodies* and *aortic bodies*, respectively. The three detected variables in this special case of chemoreception are referred to as blood gases (PCO_2, PO_2) and pH. Only the peripheral chemoreceptors respond to changes in all three variables. The central chemoreceptors respond to changes in PCO_2 and pH—not PO_2. Thus any time the blood flowing past the peripheral chemoreceptors has a change in blood gases and pH, the lungs respond by increasing either the rate and/or depth of respiration. Whenever the pH or PCO_2 of the interstitial fluids surrounding the central chemoreceptors change, the rate and depth of respiration also change accordingly.

The mechanism of changes in respiration caused by the stimulation of central and peripheral chemoreceptors involves respiratory control centers that are located in the brainstem and the sensory nerve tracts that connect these centers with the respiratory chemoreceptors. The mechanism also involves motor nerve tracts that connect the respiratory control centers to our respiratory apparatus (e.g., the diaphragm, respiratory or intercostal muscles that operate the ribs and rib cage, and respiratory conduits). Changes in blood gases and pH can either increase or decrease ventilation of the lungs. Generally speaking, decreases in pH and PO_2, and increases in PCO_2 increase ventilation. Changes in the opposite direction decrease ventilation.

Imagine that you are a resident of the mid-Atlantic region and that you live on the coast at sea level. You are adventuresome and want someday to summit Mount Everest. In preparing, you decide to climb a series of lower peaks that range from 5,000–7,000 feet (e.g., Mount Washington in New Hampshire or Clingmans Dome in Tennessee) to 12,000–15,000 feet (the Tetons, in Wyoming or one of Colorado's 14,000-foot-plus peaks). As you train, you notice that each time you ascend to and above 6,000 feet, your rate of respiration increases. A more careful inspection reveals that your depth of ventilation increases at the same time. Most inexperienced climbers (and probably even some skilled mountaineers) don't give these changes a second thought. Conversely, the investigating respiratory physiologist seeks to understand the physiological mechanism(s) that explain these changes.

As the climber ascends, he or she is continuously bathed in atmospheres that have both decreasing barometric pressures (P_B) and decreasing alveolar partial pressures of oxygen (P_AO_2). These decrements result in decreasing partial pressures of oxygen in the alveoli and the arterial blood (*hypoxia*). When the partial pressure of oxygen in the arterial blood falls below some critical level, the peripheral chemoreceptors of the carotid bodies are stimulated. This leads to increased sensory information flowing over sensory nerves to the respiratory control centers of the brainstem.

The sensory information is received and processed by central neurons that can increase both the rate and the depth of respiration. The rate of ventilation is increased so that the lungs are inflated and deflated more frequently. The depth of ventilation (i.e., increase in tidal volume, or the volume of air moving into and out of the alveoli during each respiratory cycle) is increased by the muscles of the ventilatory apparatus contracting more vigorously (e.g., the diaphragm and the respiratory muscles connecting ribs to each other and to the vertebral column).

The net effect of this is an increase in the volume of air (and hence oxygen) moved into and out of the lungs during each respiratory cycle. Another benefit of increased frequency and tidal volume is the heightened removal of carbon dioxide. As the climber ascends and works more vigorously, carbon dioxide is produced at increasing rates, thereby increasing the PCO_2 of both venous and arterial blood (*hypercapnia*, or an excess of carbon dioxide in the blood, is another stimulus for increased ventilation). Actually, elevated carbon dioxide and reduced oxygen at the peripheral chemoreceptors augment each other's effects on respiration. Thus the two stimuli act cooperatively in producing an enhanced ventilatory response that is greater than either would achieve alone—that is, hypoxia augments the ventilatory response to hypercapnia, and hypercapnia enhances the ventilatory response to hypoxia.

When there is an abundance of oxygen in the lungs (hyperoxia) or a deficit of carbon dioxide (hypocapnia), peripheral and central chemoreceptors are inhibited until the partial pressures of these two respiratory gases in arterial and venous blood return to more physiologically normal levels.

CELLULAR MECHANISMS OF RESPIRATORY
RESPONSES TO HYPOXIA AND HYPERCAPNIA

As early as 1868, German physiologists reported that hypoxia stimulates breathing. This discovery promoted a continuing investigation to identify structures that detect systemic oxygen (O_2) levels and trigger physiological responses. Nearly 50 years later, other investigators identified carotid bodies as the sensory organs that monitor the partial pressures of oxygen in arterial blood. These researchers subsequently demonstrated that hypoxia-induced stimulation of breathing results from a chemosensory reflex originating in the carotid body. Since the identification of the carotid body as an oxygen-sensing organ, much attention has been focused on uncovering the mechanisms responsible for the stimulus-response sequel. Specific details of what takes place chemically, metabolically, and physiologically inside the glomus cells of the carotid bodies are largely unknown. However, considering that the carotid body responds to hypoxemia so rapidly (within seconds), it is likely that the stimulus-response process involves changes in existing proteins rather than the synthesis of new ones, which takes minutes to hours.

Revolutionary advances in biochemistry, cell and molecular biology, and genetic engineering in the past 50–75 years have provided much insight into our understanding of physiological systems at the cellular, subcellular, and molecular levels. However, shifting research funding from the whole animal to the cellular and molecular levels has produced a marked decline in research at the physiological systems level (e.g., the cardiovascular, respiratory, and renal organ systems). At the same time, a growing fraction of the adult human population and 50 percent of premature infants experience sleep apnea (interrupted, periodic breathing during sleep) and related systemic diseases. Individuals who experience these chronic disorders frequently display serious side effects, including acid/base disturbances, hypertension, ventilatory abnormalities, myocardial infarctions (heart attacks), metabolic syndrome, and strokes. Understanding and treating these disorders has been a main focus of respiratory medicine in recent years.

Mechanistically, the discovery and elucidation of the roles of hypoxia-inducible factors in maintaining oxygen homeostasis have been key elements in the recent expansion of the study of human physiology and oxygen. For example, chronic or clinical hypoxia and experimentally prolonged hypoxia initiate, over the course of hours to days, the production and release of hypoxia-inducible factors to maintain the oxygen homeostasis. Other examples of physiological responses to hypoxia include increased production of red blood cells, formation of new blood vessels (*angiogenesis*), and the metabolic reprogramming of cells. Increased production of red blood cells (*erythropoiesis*) improves the carrying capacity of oxygen by the blood. Angiogenesis enhances the transport and exchange of oxygenated blood with tissues. Angiogenesis also brings cells in closer proximity to blood vessels. This reduces diffusion distances between the two and improves respiratory efficiency. And reorganization of metabolic machinery, including the increased production of mitochondria, optimizes the subcellular use of oxygen during periods of oxygen deprivation.

The physiological roles of recently discovered "gas messengers" are also viable mechanisms of action for maintaining the homeostasis of blood oxygen. Both nitric oxide (NO) and carbon monoxide (CO) are gases that function as *gasotransmitters*, or inhibitory transmitters, in the carotid body. Hydrogen sulfide (H_2S) is another gasotransmitter, and it functions as a stimulatory agent during hypoxia. There are several enzymes involved in the formation of all three gases, and the main enzymes are found in glomus cells of the carotid bodies. When genetically modified mice that are missing the enzymes for CO and NO production are exposed to hypoxia, the release of both gases is reduced, and the respiratory responses to hypoxia are impaired. When drugs that inhibit these enzymes are administered to mice and rats, their respiratory responses to hypoxia are also reduced.

The enzymes required for NO and CO production are heme-containing proteins, and their enzymatic activity requires molecular O_2. Like O_2, both NO and CO bind to heme. Therefore, most if not all the physiological actions of NO and CO are coupled to the activation of heme-containing proteins. For these reasons, NO and CO can be considered to resemble O_2 in some respects. Conversely, H_2S shares little in common with hypoxia. However, all three gasotransmitters exert at least some of their effects by modifying ion channels in cell membranes, including the glomus cells of the carotid bodies.

Metabolic and Respiratory Acidosis and Alkalosis

One of the most intriguing arguments, at least for this writer, for the intelligence of our bodies is the cooperativity that exists between the kidneys and the lungs in sustaining life through the homeostasis of acid-base balance. Pathophysiological changes in acid-base homeostasis can occur acutely or chronically. They can last from a few minutes or hours to years or even decades. In this regard, the respiratory system was designed in part to respond rapidly (within seconds to minutes) to changes in acid-base homeostasis. Conversely, the kidneys respond more slowly (within hours to days).

Changes in the respiratory system can compensate rapidly, whereas changes in the kidneys adjust more slowly. Several examples are worth considering.

To make intelligent use of acid-base data, one must know the limits that normal values are likely to fall within. Such published observations have been made on healthy, average persons living at sea level. These observations show that greater than 95 percent of those sampled have arterial blood-gas data that fall within the following ranges of normal: pH = 7.35–7.45; HCO_3^- = 23–28 millimoles per liter; PCO_2 = 35–48 mmHg; and PO_2 = 80–100 mmHg.

If such a normal adult were to relocate permanently to Denver, Colorado (elevation of about 5,000 feet above sea level) or Salt Lake City, Utah (elevation of about 4,500 feet above sea level), where the barometric pressure is roughly 625–650 mmHg, their resting PCO_2 after several weeks of residency would be about 32–36 mmHg. Because the partial pressure of atmospheric oxygen is lower at these elevations than at sea level, the newcomer will become modestly hypoxic upon arrival in Denver or Salt Lake City. This modest lack of oxygen will be partially compensated for by a stimulated respiration leading to mild respiratory alkalosis (elevated arterial blood pH and reduced blood PCO_2). Subsequently, the kidneys will respond to this mild respiratory alkalosis by retaining hydrogen ions and excreting bicarbonate ions in the urine. This will restore any deviations in arterial blood pH back within their normal limits while simultaneously reducing arterial bicarbonate concentrations to below-normal concentrations (e.g., fewer than 23–28 millimoles per liter). This compensatory sequence of events is called *metabolic renal compensation*, in this case for the altitude-induced respiratory alkalosis.

Converse compensatory changes are made by either the kidneys or lungs for most pathophysiological changes in acid-base homeostasis. For example, in chronic renal failure, where the kidneys lose capacity to excrete hydrogen ions and arterial blood pH falls as a consequence, a condition called *metabolic acidosis* ensues. Because of the elevations in arterial blood acidity, the respiratory system is stimulated, and excess carbon dioxide is blown off. This "respiratory alkalosis" compensates for the failed renal function and helps restore acid-base homeostasis to normal. These are good examples of the intelligence used by the kidneys, lungs, and nervous systems to solve acid-base problems created by the temporary dysfunction of either lungs or kidneys.

Clinically speaking, the data (blood samples) from which acid-base status is calculated are collected manually (by a nurse, medical assistant, etc.) and measured chemically, electronically, and/or physically. At each step there is room for human error through either inexperience and/or unintelligence. To date, all surveys of the clinical laboratories where such information has been collected reveal that even the best-trained technicians occasionally make mistakes. The worst technicians report wildly erroneous results that can lead to devastating consequences for the patient.

Summary

Chemical systems such as the carbonic acid / bicarbonate, dihydrogen phosphate / hydrogen phosphate, and ammonia / ammonium ion systems are important regulators of acid/base homeostasis in our bodies. Additionally, renal function and the respiratory system play important roles at the organ-systems level. The kidneys operate by balancing the reabsorption of bicarbonate ions and the excretion of hydrogen ions. They also regulate tubular pH by adjusting production, secretion, and excretion of dihydrogen phosphate / hydrogen phosphate and ammonia / ammonium ions. In the lungs, tidal volume and respiratory rate respond to central and peripheral chemoreceptors that are stimulated by changes in blood pH and blood gases including hypoxia and hypercapnia. Finally, gasotransmitters such as nitric oxide, carbon monoxide, and hydrogen sulfide also influence respiration. Their roles are still revealing new information about cellular and molecular mechanisms and about the ABCs of acid-base chemistry in humans and other species.

6

CARDIOVASCULAR
HEMODYNAMICS

It has been estimated that in the normally healthy adult human, no single cell is more than one micrometer away from the nearest blood supply. Simply stated, this close proximity between blood flow and cellular activity maximizes, by diffusion, the efficient delivery of oxygen and removal of waste products from the cell (e.g., carbon dioxide). One micrometer is a distance that cannot be discerned by the unaided human eye. However, we can all imagine or even measure the width of a well-sharpened pencil tip, a distance of about one millimeter that can be discerned by the healthy unaided human eye. It would take one thousand much smaller pencil tips, each with a width of one micrometer, to equal this one millimeter. One micrometer is about the distance a gas molecule must travel to get from capillary blood to the nearest cell and vice versa.

From the moment of conception, when an egg (oocyte) and sperm cell (spermatocyte) merge, the newly created fertilized egg (zygote) receives oxygen and nutrients by diffusion. Waste products are eliminated by the same process. The zygote passes through several developmental stages from conception to newly delivered infant. In order of size, from smallest to largest, these stages are as follows: zygote, conceptus, embryo, fetus, and neonate (newborn infant). There are more definitive classification schemes, and not all developmental biologists or embryologists would agree with my order. Regardless, somewhere along this path of development, the mass of multiplying cells becomes too large for diffusion to deliver oxygen and remove waste products. A problem has thus arisen in how to exchange materials; it must be solved.

At some point between conception and 18–21 days, evidence of a developing heart, circulatory system, and blood medium can be found in humans. That is, by the time the conceptus or embryo is too large to meet its metabolic needs by diffusion (the problem), a pump (the heart), conduits (blood vessels), and a transport medium (the blood) have already begun to develop (problem intelligently solved). Evidence of such developmental intelligence includes the ability of neonatologists and pediatricians to hear a heartbeat between 18–21 days postconception. Hearing this heartbeat is often the first physical evidence the pregnant woman has that she is carrying a living, developing organism.

From the third week postconception to death at any age, the cardiovascular system is charged with replacing the process of diffusion for the delivery and exchange of all blood-borne materials. If this system fails in an otherwise healthy person, a new problem is created. Imagine that the circulatory system develops a leak (problem: unwanted loss of blood) or that unexpected side effects of a drug, prescribed or otherwise, weaken the heart as a pump (another problem: reduced circulation). Similarly, the heart's failure to generate and/or conduct a pacemaking action potential and the rhythm disturbances this might cause is a third, electrical problem. All of these and many more problems must be solved by the cardiovascular system.

One purpose of this chapter is to provide several cardiovascular examples of the human body's intelligent solution to such problems.

As a starting example, consider a ventricular premature beat (VPB). This is a condition where the ventricles relax and contract aphysiologically during a single cardiac cycle. Usually, this happens before the ventricular chambers have fully filled with blood. Therefore, when they contract subsequently, they eject a reduced volume of blood and the circulatory system, during that premature cardiac cycle, is compromised. If this VPB replicates itself many times or if the rhythm disturbance continues unchecked, the end result can be circulatory collapse and death.

All of us experience ventricular premature beats. They occur randomly during both waking and sleeping hours. They are not considered a major health problem to either the cardiovascular system or to the individual as a whole. More serious ventricular arrhythmias such as ventricular salvos (VS) and ventricular tachycardia (VT), however, can be of grave concern and even life-threatening. When a VPB occurs, commonly the very next normal ventricular beat and its corresponding ejection of blood (i.e., the stroke volume) take place only after a compensatory pause. This can be recorded experimentally and displayed on a computer monitor; that is, the characteristics of the cardiac cycle during and following a compensatory pause can be quantified.

The compensatory pause provides more time for the ventricles to fill with blood. The increased filling compensates for the reduced filling during the previous cardiac cycle. Increased filling and the more vigorous contraction it subsequently causes happen via the Starling effect. The Starling effect results from excess stretching of the heart. When any cardiac chamber, whether atria or ventricles, is filled with too much blood, it is stretched excessively. The excessive stretching causes the muscle cells of the chamber to contract more forcefully. This added force of contraction expels the extra volume of blood that was the initial cause of excessive stretching. After a few more cardiac cycles, the excess blood is ejected, and the volume and stretch of the chamber are restored to their prior physiological states.

The Starling effect can be reproduced and quantified with the proper recording system in both an experimental laboratory and in the clinic. In my teaching physiology laboratory at Rutgers University, we demonstrate the Starling effect using in situ frog hearts (hearts that are still inside the bodies of euthanized frogs). The apex of the heart's ventricle is impaled with a fine, stainless steel hook. The opposite end of the hook is attached to a suture that is attached to a force transducer. The force transducer is part of a micromanipulator apparatus that can be moved up or down in 1-millimeter increments. Such movement stretches or relaxes the heart in quantifiable units of length. This causes changes in the contractile behavior of the heart. Such an experiment can be managed using a modern data acquisition system composed of an A-D converter (to convert analog signals to digital) attached to a desktop computer running software suitable to the experiment. Signals can be amplified, filtered, and recorded by the computer.

After baseline data for heart rate and contraction are collected, the heart is stretched incrementally by two to three millimeters. The force of contraction associated with each increment is recorded. Subsequently, the heart is returned to baseline conditions and allowed time to achieve a new steady state. Students quantify the effect of stretch on the force of contraction. The greater the amount of stretch applied, the greater the subsequent force of contraction. This experiment is a wonderful demonstration of what we call the Frank-Starling law of the heart, and it is an intelligent solution to the problem of reduced ejection volume associated with ventricular premature beats.

The excess filling of ventricles with blood following a ventricular premature beat and its corresponding compensatory pause ensures that total blood flow to the body is not reduced. Total blood flow is called *cardiac output* (expressed in units of liters/minute, or L/min), and there are redundant backup systems built into our intelligent cardiovascular systems to protect it. Thus the amount of blood flowing to the entire body during a one-minute period of time does not change significantly despite a VPB and its reduced stroke volume. The Starling effect occurs because it is preceded by a compensatory pause resulting in greater filling of the ventricles subsequently (problem solved).

STARLING FORCES AND EDEMA/INFLAMMATION/SWELLING

Another wonderful example of cardiovascular intelligence that was originally described by Ernest Starling, a 19th-century British physiologist, is the application of Starling forces to the homeostasis of blood volume and body water. In this case, Starling's research was focused on the microcirculation and exchange of fluids and materials transmurally (across the walls of capillaries and postcapillary venules) with the interstitium, or extracellular spaces. His objective was to identify and quantify the factors that contribute to the movement of fluids and proteins out of the microcirculation and into interstitial spaces. We have come to call these factors *Starling forces*.

The reason this phenomenon is important is because Starling forces help sustain circulating blood volume and other body compartments that contain water and its salt solutions. The average healthy adult has a circulating blood volume of about 5.0 liters. This volume is circulated by the heart and vasculature each minute. The 5.0 liters are distributed equitably under baseline control conditions to the various organs and tissues of the body according to their metabolic, neurogenic, and physiological needs. Each organ and tissue receives a well-quantified fraction of the 5.0 liters. For example, the brain receives about 15 and the heart about 5 percent of the total cardiac blood output each minute. The gut and kidneys receive the lion's share, at 25 and 20 percent, respectively. These numbers can and do change—for example, during sustained heavy exercise, with hemorrhaging following an automobile accident, and so on. There are four Starling forces, and they are all expressed in units of pressure (i.e., mmHg, or millimeters of mercury): capillary hydrostatic pressure, interstitial hydrostatic pressure, capillary oncotic pressure, and interstitial oncotic pressure. Hydrostatic pressure is caused by the volume of fluid inside the two compartments (also called *intravascular blood volume* and *interstitial* or *extravascular fluid volume*) and by the physical properties of the compartments (e.g., the compliance of blood vessels). Oncotic pressure (also called *colloid oncotic pressure, colloid osmotic pressure,* or *crystalloid oncotic pressure*) is caused by the concentration of proteins inside the two compartments. The important consideration is the net effect that one or more of these four variables has on the direction and movement of fluid into or out of the two compartments (i.e., the intravascular and extravascular, or interstitial, compartments).

Capillary oncotic and interstitial hydrostatic pressures were designed to keep fluid inside the vascular compartment. Interstitial oncotic and capillary hydrostatic pressures cause fluid to move out of the vascular space and into the extravascular space. The continuous movement of blood through the circulatory system and of interstitial fluid through extracellular spaces is necessary to sustain the health of the body. Fluid movement also maintains homeostasis of a wide range of physiological processes (e.g., lymph flow in the lymphatic circulatory system and the protein, electrolyte, and cellular constituents of the lymph nodes). Therefore, the importance of Starling forces to the healthy maintenance of life cannot be overstated.

Edema and inflammation are two cases in point. Actually, they are really one case in point—what we so commonly call "inflammation" in the world of health, medicine, and misinformation about both, is really edema. Fields such as cell and molecular biology and immunology have mushroomed explosively in the past several decades. Because investigators and clinicians in these fields have used the term *inflammation* so loosely and without fundamental knowledge of Starling forces (most such clinicians and scientists have little or no background in classic mammalian physiology), much confusion about the topic has arisen. For example, consider television and internet advertisements and confusion about when to apply hot versus cold compacts to sports-related and other injuries. (Do we alternate between hot and cold . . . or is that cold and hot?)

Edema (not *angioedema*; all edema results from blood vessels, and *angio-* is a prefix meaning "blood vessel") results from imbalance in Starling forces. So does inflammation. Here is a brief list of factors that are likely to cause edema, all of them representing changes in the homeostasis of Starling forces: (1) increased permeability or porosity of capillaries and postcapillary venules, (2) reduced lymphatic circulation, (3) excess accumulation of proteins in the extravascular spaces, (4) reduced uptake of proteins and water by lymphatic circulation, (5) loss of water from systemic capillaries at a greater rate than uptake by lymphatic capillaries, and (6) any combination of the foregoing.

One or two examples will help drive the message home. If you are sufficiently sensitive to insect venom and a wasp stings you on the wrist, chances are that your forearm (and perhaps the upper arm and other body parts as well) will soon swell. The swelling is caused by edema. Edema and swelling are caused, in this case, by the rapid loss of water and proteins from the intravascular compartment (*extravasation*) and by the concurrent accumulation of water and proteins in the extravascular spaces. The ability of the lymphatic system to drain the excess water and proteins and to keep pace with their rates of loss from the systemic circulation at the same time is temporarily overwhelmed.

Mechanistically, toxins and allergens in wasp venom stimulate release of histamine, bradykinin, and other compounds from enterochromaffin and mast cells distributed throughout human tissues. These biogenic amines and other cytokines increase permeability and porosity of systemic capillaries and postcapillary venules. Biogenic amines and cytokines are groups of chemicals in the blood or tissues surrounding blood vessels. They can increase the loss of water and other elements from the blood and into the tissue spaces. They aggravate this action by also being vasoactive, thus causing changes in upstream and downstream vascular resistances that are conducive to edemogenesis. In short, the homeostasis of capillary Starlings forces has been disrupted by the wasp venom.

As a second example, I will use my personal experience. While writing this chapter, I had orthoscopic surgery on my right knee to trim torn cartilage. I have had similar surgery on both knees in the previous 10–25 years. In this latest experience, and within 24–48 hours postsurgery, my right leg from the knee to the foot became increasingly edematous. The tissues swelled rapidly, visibly, and painfully (this had not happened with any of the prior surgeries). To treat the condition, I applied ice packs (as time and circumstances allowed) for 45–60 minutes two to three times per day. As much as was possible, I also kept my foot elevated well above my heart. It took applying this methodology rigorously and daily for two or more weeks to gain control of the edema and to begin seeing signs of reduced volume in my leg.

I am still uncertain of what caused the edema, and I am confident that neither the orthopedist nor the anesthesiologist would have known either. However, after receiving the written reports from both, my personal physician (a person to whom I had taught physiology some 20 years earlier at Rutgers University) and I reasoned that a combination of the medications used by the anesthesiologist and the surgical procedure were the causes. For example, the anesthesiologist used a serotonin inhibitor to reduce the prospects of me experiencing postsurgical nausea (something he did without my knowledge and probably without his knowing that serotonin influences Starling forces and can cause edema in skin and skeletal muscle).

In these examples—a wasp sting and the drug cocktail used in orthoscopic surgery—the swelling can be explained in terms of disrupted Starling forces, edema, and little else. There is no reason to apply the designation of *inflammation* to these injuries. The swelling has a completely sensible pathophysiological explanation: edema. In both cases, the same problems arose of disrupted Starling forces and accumulation of fluid in the interstitial spaces. The cardiovascular system subsequently solved these problems by restoring the original rates of fluid loss from the capillaries and postcapillary venules and by increasing drainage of the interstitium by the lymphatic circulation system.

What about the application of hot or cold packs to these injuries? In both cases, cold packs will do; no hot applications are warranted or needed. One should apply cold packs as soon as possible after the injuries occur and keep them in place for about 45–60 minutes (ideally two to three times per day). Also, keep the affected limb or joint elevated well above the heart while the cold packs are in place. Elevation of the swollen joint enhances the flow of venous blood out of the joint and reduces the volume of blood inside the veins. This reduces capillary hydrostatic pressure in the affected area's microcirculation, which in turn reduces the rate of fluid loss into the interstitium.

The logic of cold versus hot applications is also based on Starling forces and the architecture of the walls of pre- and postcapillary blood vessels. Upstream to all capillaries are the *resistance arterioles*. They are given this name, in part, because of their thick, muscular walls (vascular smooth muscle). Postcapillary venules also have vascular smooth muscle in their walls, but the smooth muscle cells are not as dense as those found in upstream resistance arterioles. Also, the ratio of vascular smooth muscle cells to other components of the wall is much greater in arterioles than in venules. All vascular smooth muscle contracts in response to cold and relaxes in response to heat. A cold compact will contract both arterioles and venules. However, the contraction and attending vasoconstriction will be much greater upstream to the capillaries than downstream. The cold will reduce the volume of blood inside the capillaries (therefore lowering capillary hydrostatic pressure). This will result in reduced loss of plasma water into the interstitium. Lymphatics will be more effective in draining the excess accumulation of water, and tissue volume (edema) will decrease.

Applying heat to such injuries will only exacerbate the problem by vasodilating upstream arterioles more than downstream venules (i.e., blood flow into corresponding capillaries will increase). With heat, capillary hydrostatic pressure will increase and so will the rates of extravasation of capillary fluid and tissue swelling.

Alternating between hot and cold therapy is also not called for in our wasp sting, sports-related joint injury, or orthoscopy examples. The only explanation this physiologist can come to of all the television ads and commercials to the contrary is the lack of knowledge and understanding of the manufacturers, advertisers, and physicians subscribing to such mysticism. (Application of heat does have a role in some cases, however. These include the strain of a muscle or muscle groups where obvious edema has not occurred. Under such conditions, heat will help relax strained muscle and other soft tissue. It will also improve blood flow, which should speed repair and recovery.)

PACEMAKERS AND CONDUCTION SYSTEMS

The pace car, or "pacemaker," in the Indianapolis 500 and other such races is a car positioned in front of the competing racers. In 2017, the pace car was a white 2017 Chevrolet Corvette Grand Sport. The pace car has two main purposes. First, it leads the assembled pack of competing race cars around the track for a predetermined number of unscored warm-up laps. If officials approve, the pace car releases the field at a purposeful speed to start the race. Second, during yellow-flag cautionary periods, the pacemaker enters the track and picks up as lead car, causing the race cars to converge and reduce their speed until the safety concern has been resolved. Similar functions are performed by state police cruisers on the busy and often congested superhighways of the northeastern United States (e.g., I-95 and the New Jersey Turnpike).

Our hearts also have a pacemaker (like the Chevy Corvette Grand Sport) and a circuit over which action potentials travel (like the Indianapolis Motor Speedway). The cardiac pacemaker sets the rate and rhythmicity of the heart. It determines how many action potentials pass over the specialized conduction system (the electrical circuit) each minute. The outcome we call *heart rate* (HR), expressed in beats per minute (bpm) or cycles per minute (cpm). The average adult resting heart rate is about 72 cycles per minute. This means the heart's pacemaker—the sinoatrial, or SA, node—generates 72 action potentials each minute. Heart rate can increase, decrease, or remain the same over many minutes or hours each day. An event that excites or stimulates us might increase our heart rate. Sleeping or relaxing quietly and motionlessly normally reduces our heart rate. During waking hours, and as long as our body position and level of activity remain relatively constant, heart rate will remain at or near 72 cycles per minute.

During competitive events, at least for most athletes, heart rate will increase substantially during the event. For example, swimmers, ice skaters, runners, skiers, and others all experience increased heart rates during competitive events. In well-trained young athletes, heart rate can increase to more than 200 cycles per minute during the peak of an event. Under baseline resting conditions, these same athletes have reduced resting heart rates; something we call *exercise-induced bradycardia*. It has been reported that at the peak of his conditioning, Lance Armstrong had a resting heart rate below 40 cycles per minute. I have a colleague who is a competitive marathoner. He runs 4,500–5,000 miles per year to compete in two or three marathons. One of his current goals is to hold the U.S. record for his age group. His basal resting heart rate varies between 35–40 beats per minute.

Other events that change heart rate are unplanned and unexpected. For example, an unexpected explosion of a subterranean gas line after midnight can cause a "startle response" in almost anyone. Such responses are associated with marked increments in heart rate that are sustained for minutes. Alternatively, any event that increases the activity of the vagus nerves innervating the heart can markedly decrease heart rate. If the event is sufficient, unexpected activation of the cardiac vagus nerves can even stop the heart momentarily (cardiac vagal arrest).

Typically, running marathons, being startled, and any other planned or unplanned cardiovascular events will be accompanied by changes outside the heart. For example, blood pressure can rise or fall unexpectedly. Blood flow, either to an individual organ or to the entire body, can increase or decrease as well. These hemodynamic changes are usually monitored by sensory receptors strategically located throughout the cardiovascular system. Changes in the normal physiology detected by such sensors are fed to the brain and central nervous system. There they enact corresponding changes that compensate for and/or reverse the initiating stimulus. In the end, homeostasis of heart rate, blood flow, blood pressure, and other cardiovascular variables is restored.

Another example should help drive the message home. Consider that you were peacefully sleeping when a gas line explosion rocked the neighborhood, awakened you, and initiated a startle response. Your heart raced as you leaped out of bed and ran to the window to investigate. As local police and fire departments arrived on the scene, your heart continued to race, and you nervously paced the floor. In addition to an elevated heart rate, both blood flow (cardiac output) and blood pressure rose to aphysiological levels. If sustained, none of these changes were good for your health.

Fortunately, your cardiovascular system has a set of checks and balances in place. One of particular importance is called the *carotid sinus baroreceptor feedback mechanism*. Upon being startled, your elevated blood pressure stimulates nerve endings in the wall of the carotid sinus. The carotid sinus is located at the division of the main carotid artery into external and internal branches (near the tip of the ear lobe). The stretch of these mechanoreceptors by your increased blood pressure was an activating signal. A volley of action potentials was carried over the carotid sinus nerve into the brainstem. The brainstem is composed of the pons and medulla. Within the medulla are found cardiovascular control centers, or groups of like-behaving neurons that regulate heart rate and blood pressure. The arriving volley of action potentials (signals) was analyzed by these centers, and efferent motor signals were simultaneously sent to various activators (cells, tissues, and organs) throughout the body.

The main activators of the carotid sinus baroreceptor system are the heart and blood vessels. Other activators include higher brain centers and the adrenal glands. Some signals cause the slowing of heart rate. Others cause different blood vessels to relax (e.g., large veins and resistance arterioles). In your case (the startle response), the net effects of the above actions were reduced heart rate, reduced blood flow, and reduced blood pressure. Despite the commotion, and because of your intelligent cardiovascular feedback control system, your heart rate and excited cardiovascular system began to restabilize shortly after the explosion. Heart rate returned to a healthier 70–80 beats per minute, and cardiac output (blood flow) fell to five to six liters per minute. Soon thereafter, systemic arterial blood pressure was restored to near 120/80 mmHg, and you were able to relax as the anxiety of the moment passed.

Regulation of Cerebral and Coronary Blood Flow

I involved a group of volunteers in an experiment one time. The objective of the experiment was to determine if caffeine has potentially harmful cardiovascular effects. Volunteers were asked to bring their favorite caffeinated beverage to the lab. They were responsible for knowing the caffeine content of the drink (e.g., 250 milligrams). Volunteers were briefly outfitted with several physiological transducers while lying in the supine position on an examination table. Lights were dimmed, and data (e.g., heart rate, blood flow, blood pressure, etc.) were collected at 5-minute intervals for 15 or 20 minutes. After collecting baseline control data, volunteers ingested their drinks in about 2–3 minutes. They resumed the supine position, and additional sets of data were collected at 15-minute intervals for about one hour.

Before designing that experiment, I read the published literature. Most of what I found were epidemiological and meta-analyses. These reports did not interest me. There were, however, some useful clinical experiments looking at the effects of caffeine on blood flow and function in the brain and heart. From these reports, and from more basic scientific literature, I gleaned valuable information about caffeine and its analogs (e.g., theophylline). For example, caffeine and its analogs are competitive antagonists of adenosine receptors, inhibitors of the phosphodiesterase family of enzymes, and manipulators of potassium channels. This won't mean much to the average reader. However, to the trained reader, they are red flags. For one thing, via its ability to bind to adenosine receptors on cerebral and coronary blood vessels, caffeine can express vasoconstrictor properties and thereby reduce blood flow.

In our experiment, we found evidence that caffeine is indeed a peripheral vasoconstrictor (probably through competitive inhibition of adenosine receptors). We also saw signs of disturbed electrical function in the heart: several volunteers who consumed the greatest amounts of caffeine daily had elevated T waves in their electrocardiograms (ECGs). Elevated T waves can reflect disturbances in potassium homeostasis, resulting in, for example, increased circulating concentrations of potassium in the blood plasma (although not always). In the average healthy adult, circulating plasma concentrations of potassium should be about 4.5 mEq/L (milliequivalents per liter). Increases to 6–8 mEq/L can cause cardiac arrhythmias. Concentrations of greater than 10–12 mEq/L can arrest the heart.

One of the main characteristics of solutions used to lethally inject criminals is hyperkalemia (elevated concentrations of potassium). Potassium is added to the cocktail because of hyperkalemia's well-established and reliable ability to produce cardiac arrest in seconds. We observed other disturbances in the ECG as well, but the most commonly seen result was peripheral vasoconstriction. Within 30–45 minutes after consuming caffeine (e.g., 150–250 mg) almost all volunteers showed signs of reduced blood flow to their fingertips.

Our sample of volunteers included a few conditioned athletes who regularly ingested caffeine capsules (e.g., 250 mg) before weight training. One mentioned that by the end of his workouts, his fingers and toes were often cold and that he felt numbness and tingling in his fingers on occasion. The opposite effects should have occurred (his fingers should have been warm as evidence of the body's attempt to lose heat). During such workouts, body temperature often rises. Most of us lose heat by sweating while working out. We also lose heat by peripheral vasodilation. These actions help keep a person's core body temperature from rising to dangerous levels.

Another volunteer asked if he could bring his caffeine capsules to the lab in place of a caffeinated beverage. I told him yes, as long as the quantity of caffeine in the capsule had been quantified. He brought a bottle of such capsules. The bottle and each capsule were labeled as containing 250 milligrams of caffeine. Out of curiosity, he wanted to open the capsule and weigh its contents before ingesting them. I have delicate analytical balances in my lab, and so we did this. Before the capsule was half empty, the scale registered 250 milligrams. This was a surprise. The capsule either contained more than 250 milligrams of caffeine or an unidentified filler.

Caffeine is water soluble, so we added the 250 milligrams we had just weighed out to tap water. The powderlike material just sat on top of the water; it did not dissolve. I added a little heat and agitation, as this often helps in such cases. The substance was still insoluble in water—a second surprise. I told my volunteer that I personally would not take such an unknown compound. He said he had been using these capsules for a couple years, and then he consumed the drink. We did the experiment even though we did not know what he had just ingested. We could not draw any conclusions about caffeine and his cardiovascular system.

Whole blood is composed of cellular and noncellular elements. Cellular elements include red blood cells (RBCs), which are the most abundant cellular elements in whole blood; white blood cells (WBCs) called *leukocytes*; and nonnuclear cell fragments called *platelets* or *thrombocytes*. In addition to these cellular elements, whole blood contains inorganic and organic chemicals in great abundance. All of these play significant roles in the health of the body. They are all suspended in a liquid matrix called *plasma*. Removing one of the organic molecules, fibrinogen, transforms plasma to serum. Because of fibrinogen, a standing sample of plasma will form a clot. Serum will not clot (because of the absence of fibrinogen).

One of the great challenges to the homeostasis of whole blood (called *hemostasis*) is for whole blood to remain thin enough to circulate (nonviscous) but thick enough to prevent escape across the vessel walls (a cause of hypovolemia, or reduced blood volume or hemorrhage). This is indeed a potential problem, and one in which the two prospects seem mutually exclusive (i.e., increase viscosity, and blood will not circulate; decrease viscosity, and we potentially bleed to death). All three classes of blood cells play important roles in hemostasis.

Platelets are cell fragments. They do not have nuclei and thus are not self-replicating. Platelets arise from another, much larger cell type called a *megakaryocyte*. This transformation occurs in the bone marrow. An individual megakaryocyte can give rise to as many as several thousand platelets. There is an important feedback mechanism that controls the production and release of platelets. It involves a circulating compound called *thrombopoietin* (TPO), a glycoprotein produced by the kidneys and liver and released into the circulating plasma, TPO receptors found on the surface of platelets, and the bone marrow.

In conditions of hypoplastic bone marrow (too little production of megakaryocytes), the circulating content of platelets declines. With fewer platelets, circulating concentrations of TPO increase because of the scarcity of platelet TPO receptors to bind the compound. The elevated concentrations of TPO stimulate the hypoplastic bone marrow, which increases its production of megakaryocytes and thus platelets. The newly released platelets increase in the plasma and bind some of the excess TPO, and the stimulus to the bone marrow production of megakaryocytes is reduced. A new steady state is smartly achieved.

There are other players in this TPO-platelet-bone marrow mechanism (e.g., IL-3, or interleukin-3—a cytokine produced by the kidneys and liver). However, the important point is this: a problem arose (hypoplastic bone marrow and reduced production of megakaryocytes and platelets), and the TPO/platelet negative feedback system intelligently compensated for it. In order for this to have happened, communication between TPO and its corresponding platelet receptors (and between the bone marrow and TPO) had to occur. Communication cannot occur without chemicals and their receptors recognizing one another. When this flow of information takes place effectively and efficiently, the problem gets solved.

Continuing my explanation of this system, circulating platelets plug small injuries in the capillary endothelium, which if left unchecked, could lead to serious loss of blood and even death. This process includes the adhesion of platelets to the site of injury, subsequent activation of the adhered platelets, and final aggregation of other platelets and cellular elements. Platelet adhesion occurs only when shear forces in and around the platelets and endothelial cells increase and in response to vessel injury and release of selected humoral factors from platelets and the endothelium. The total process can be thought of as the triple As: adhesion, activation, and aggregation. It is important to note that under physiological conditions, circulating platelets do not adhere to themselves, to other circulating cellular elements, or to the vascular endothelium. It would be disastrously unintelligent if they did. One preventive measure is the negatively charged surface areas of both platelets and vascular endothelial cells, which cause them to repel one another under normal physiological conditions.

Instead, the joining of platelets to themselves and/or to the vascular components (adhesion) is mediated by platelet receptors. A breach in the vessel wall exposes these receptors to binding agents (ligands) that are integral elements of the matrix that helps hold endothelial cells together to form the walls of capillaries and other vessels. Some of the best-known ligands are collagen, fibronectin, and laminin. One ligand that occurs naturally in blood plasma is von Willebrand factor (VWF). This ligand is made by endothelial cells and megakaryocytes (thus it is found also in platelets). Some of the triggers for release of VWF are high shear, selected cytokines, and tissue hypoxia.

Binding of these tissue ligands triggers a change in the conformation, or structure, of the platelet. The change in structure initiates an intracellular signaling cascade. This leads to an exocytotic event known as the *release reaction* or *platelet activation*, which involves the release of the platelet's contents. Activated platelets release the contents of storage granules (vesicles) such as adenosine triphosphate (ATP), adenosine diphosphate (ADP), serotonin, calcium ions, VWF, clotting factor V, and fibrinogen. Platelet activation also leads to the metamorphosis of the cell from a disc-like structure to one with many fingerlike projections called *filopodia.*

Signaling molecules released by the activated platelets and from injured tissues amplify the platelet activation response. Adenosine diphosphate, serotonin, and one other signaling molecule, thromboxane A_2, activate additional platelets, and this promotes platelet aggregation. Aspirin, which blocks the enzyme cyclooxygenase, inhibits clotting by reducing the release of thromboxane A_2. Fibrinogen, which is always present in circulating blood, forms a bridge between adjacent platelets and thus participates in the creation of a platelet plug. The platelet plug eventually stops blood loss through the damaged endothelium.

The history of blood clotting begins with the Finnish physician Erik Adolph von Willebrand. The son of a district engineer, von Willebrand earned his medical degree in 1896 at the University of Helsinki. He then earned a second doctoral degree by studying the changes in blood after a hemorrhage. For the remainder of his career, the properties of blood and its coagulation were the focus of von Willebrand's research and professional interest.

Von Willebrand was the first to describe the blood coagulation disorder later named for him (von Willebrand disease). The event that caught his interest was the case of a five-year-old girl from the Åland Islands (off the coast of Finland) whose family had an extensive history of bleeding. The girl was the 9th of 12 children in the family. Four older siblings had bled to death at early ages, and Hjordis, the five-year-old index case, would later die during one of her early menstrual cycles. Mapping the family history, von Willebrand found that 23 of the girl's 66 extended family members were affected. The disease was more common among women in the family.

The study of hemostasis has made much progress in the nearly 100 years since von Willebrand's early work. We now know that the coagulation cascade beginning with adhesion, activation, and aggregation involves both extrinsic and intrinsic pathways. These come together with the conversion of prothrombin to thrombin to form a common pathway that involves fibrinogen, fibrin polymers, and the formation of a stable fibrin clot. More work is needed to continue unraveling the mysteries of hemostasis and congenital accidental blood loss.

CHEMICAL CONTROL OF BLOOD FLOW
AND OXYGEN HOMEOSTASIS

Earlier I mentioned the effects of caffeine on the cardiovascular and other systems. In this section, I will describe the role of an important, naturally occurring chemical that helps regulate coronary blood flow as well as how caffeine and similar compounds pose a potential problem for this regulatory system. I should say at the outset that what I discuss here pertains not only to coronary circulation and the heart but also to the brain and its cerebral circulation (as well as to many other organs and their blood supplies).

When physiological tissue oxygenation fails or decreases, the net production of high energy compounds that prevails during normoxia (physiologically normal oxygenation of tissues) transforms to net energy degradation during hypoxia. During normoxia, ATP is produced. During hypoxia, ATP is dephosphorylated to form the following, sequentially: ADP, adenosine monophosphate (AMP), and adenosine (ADO). This process occurs ubiquitously in human (and other animals') organs and tissues during periods of oxygen deprivation.

Adenosine is a well-known vasodilator. Its vasodilatory properties have been most thoroughly investigated in the heart, the brain, and in skeletal muscle. Adenosine's importance as a coronary vaso-dilator was first reported in the early 1960s by Robert M. Berne at the University of Virginia. Berne's adenosine hypothesis for the chemical regulation of blood flow states that whenever tissues experience hypoxia, production of adenosine increases. The newly formed adenosine vasodilates blood vessels, blood flow (hence oxygen delivery) increases, and hypoxic conditions are replaced by a new normoxia. During the 1970s, I began doing my own experiments with adenosine and coronary circulation.

In one of my early experiments, we studied the adenosine hypothesis in isolated guinea pig hearts. After isolating and instrumenting these hearts, we allowed several minutes for the monitored variables to achieve physiological steady states (e.g., spontaneous heart rate and coronary blood flow). Then we administered incremental doses of adenosine and monitored corresponding increments in coronary blood flow. Two or three doses were usually sufficient to produce a doubling, tripling, or quadrupling of flow. After administering adenosine, we allowed several more minutes for it to wash out of the tissues and for blood flow to reachieve the preadenosine, baseline steady state. Subsequently, we added theophylline, an analog of caffeine and an adenosine receptor blocker. After several minutes, once monitored variables were in their post-theophylline steady states, we readministered the adenosine doses. Under these conditions, the coronary blood flow response to the adenosine was significantly attenuated. Adenosine did not have the same vasodilatory efficacy in the presence of theophylline as it did in its absence. These results were reproduced, written up for peer review, and subsequently published in the scientific literature. From those and related experiments, we learned that the physiological actions of adenosine are reduced in the presence of methylxanthines (theophylline, caffeine, and related compounds).

Ten years later, we repeated similar experiments in anesthetized, instrumented dogs. After baseline data were collected under normoxic conditions, we ventilated the dogs on mixtures of gases with low oxygen contents (i.e., we made the dogs hypoxic). Coronary blood flow increased three- to fourfold under these conditions, just like it did in isolated guinea pig hearts during hypoxia and with administered adenosine. After a first exposure to hypoxia, dogs were reventilated on room air and allowed to reachieve normoxic conditions. Subsequently, we added either another adenosine receptor blocker (8-phenyltheophylline) or the naturally occurring enzyme adenosine deaminase, which degrades adenosine to the physiologically inactive inosine. A second exposure to hypoxia under these experimental conditions was accompanied by significantly attenuated coronary blood flow responses. By degrading adenosine, the enzyme made less of the nucleoside available to bind to its coronary vascular receptors. Or by occupying the adenosine receptors with 8-phenyltheophylline, naturally occurring adenosine that had been released in excess during hypoxia could not bind to them. In both cases, the same outcome resulted: reduced delivery of oxygen to a needy heart.

Oxygen supply (via coronary, cerebral, or skeletal blood flow) and oxygen demand in any single cell, tissue, or organ system must remain in balance in order for the cell, tissue, organ, or animal to survive. Oxygen imbalance and oxygen debts occur when the demand for oxygen is greater than the corresponding supply of oxygen. If a problem-solving challenge requires greater flow of blood to a region of the brain, and if adenosine is the mediator of the potential increase in cerebral blood flow, then the need for additional flow (more delivery of oxygen) cannot be met if the vascular adenosine receptors are occupied by an adenosine receptor antagonist (e.g., caffeine, theophylline, 8-phenyltheophylline).

Extraction of more oxygen is an alternative mechanism an organ can use to get additional oxygen. Supply of oxygen is the product of upstream (arterial) blood flow and the content of oxygen in that arterial blood. Extraction of oxygen is determined by subtracting the downstream (venous) oxygen content from the arterial oxygen content. The difference represents the amount of oxygen the tissues extract from each unit of blood during a defined period of time (usually one minute). If an increased demand for oxygen is placed on the circulation, most tissues and organs can get the additional oxygen via one of two mechanisms. First, they can increase their blood flow (vasodilation). This will supply more oxygen per unit of time. Second, they can extract more oxygen from the arterial blood supply, leaving less oxygen in the downstream, venous blood.

The brain and skeletal muscles are examples of organs that have the above choice between more flow and greater extraction. The heart is an example of a more metabolically active organ that does not have a choice. Even under baseline resting conditions, the mammalian heart extracts about 75–80 percent of the arterial oxygen made available to it. Thus if an excess demand for more oxygen is made, the heart's only option is to get it by increasing coronary blood flow with vasodilation. Since adenosine is one of the main and most important naturally occurring coronary vasodilators, its value to the heart cannot be overemphasized.

As publicized elsewhere, it is estimated that greater than 90 percent of the adult population in the United States regularly consumes caffeinated beverages. The bulk of this caffeine consumption comes from coffee even though there is a growing trend to also consume large quantities of energy drinks that are heavily caffeinated. Young adult college students who believe they need caffeine to stay up late cramming for exams is just one example of a foolish and unnecessary problem badly in need of being solved intelligently.

Imagine a young college student has just consumed his latest cup of coffee or energy drink at 1 or 2 a.m. Also imagine that this drink contained the maximum content of caffeine. Now imagine that a dorm fire forces the student to rapidly descend several flights of stairs (since elevators are off-limits during fires and fire drills). An increased demand for oxygen by the heart ensues. The heart was already extracting maximum oxygen while the student was studying prior to the fire alarm. Now the heart's only option for more oxygen is coronary vasodilation, a mechanism largely dependent on adenosine. Unfortunately, this student's adenosine receptors are already occupied by a competitive antagonist—caffeine. The heart is unable to get the additional oxygen it needs, and the student's safety and health are at serious risk.

Cardiovascular diseases, including hypertension and stroke, still cause more deaths and disabilities in industrialized nations (such as the United States) than many other causes of death combined. Despite reduced federal funding for this field in recent decades, not all the important questions have been answered. For example, To what extent do chemotherapy and radiation treatment for cancer damage the heart and circulation? This is a question for which there are few if any definitive answers, because there has been no funding for research to answer it. Still, our hearts go on beating minute after minute and year after year despite the challenges and problems we voluntarily impose on them. Sedentary lifestyles, unhealthy diets, and the use of tobacco, alcohol, caffeine, and other drugs all reveal how little gratitude we possess for such an intelligent and marvelous organ system.

7

RESPIRATION AND ITS CONTROL

One of my research interests is the physiology of hypoxia. *Hypoxia* means different things to different people. To me it means an inadequate supply of oxygen to the body. Inadequate supply means insufficient quantities of oxygen to sustain physiological function at any level (e.g., organ system, tissue, cell, and subcellular levels). I have investigated and taught about the phenomenon of hypoxia at different levels for many years.

I was born and raised in the mountains and valleys of the intermountain West, called the Rocky Mountains However, I have spent nearly the last 50 years living at or near sea level in places such as central New Jersey, New Orleans, and East Lansing, Michigan. The hospital where I was born in Afton, Wyoming, is at an elevation of nearly 6,000 feet. The hospitals where my children were born in central New Jersey are at about 100 feet. The summit of Mount Everest, the highest point on earth, is 29,028 feet above sea level.

In these locations and in all others on the surface of the earth, we are bathed in a sea of gases. These gases can be divided into two categories: respiratory and nonrespiratory. Respiratory gases are oxygen, carbon dioxide, water vapor, and nitrogen (though the latter two might be contested by some readers). Nonrespiratory gases include the lesser-known argon, helium, krypton, neon, radioactive radon, and xenon (noble gases). When it comes to respiration, most of us don't give a second thought to the noble gases.

When respiratory physiologists speak of inspired and expired gases, they only include carbon dioxide, nitrogen, oxygen, and water vapor. This is true no matter if the topic of discussion is respiratory mechanics, exchange and transport of gases, control of respiration, or the gas equations and laws. Only about 75 percent of a breath of freshly inspired air becomes available to the alveoli for exchange with the pulmonary blood. This is because of the volume of used air that fills the conducting airways and respiratory passages before fresh air arrives. *Used air* means air that has already been exchanged with pulmonary blood and has a low content of oxygen and a high content of carbon dioxide. The conducting airways are also known as *dead space*. When a person exhales, about 25 percent of the used air is not exhausted to the atmosphere. Instead, it remains in the conducting airways, filling the dead space volume. Thus during the very next inspiration, the volume of dead space air is the first to be redelivered to the alveoli.

The average healthy young adult inspires a resting tidal volume of about 500 milliliters (approximately one pint) of air with each breath. In other words, tidal volume is the amount of air passing into and out of the lungs during a single inspiration or expiration. This volume takes its name from the waxing and waning of ocean tides. The same pattern is followed by ocean waters washing onto and away from the shore with each tide. Similarly, this rhythm is followed by air flowing into and out of the lungs during each respiratory cycle. The volume of the respiratory passages is about one quarter of the tidal volume.

Dead space air, although not available for gas exchange, does serve a useful purpose. By partially filling the conducting airways during expiration, dead space air ensures that they do not completely collapse before the onset of the next inspiration. This makes it easier for the subsequent volume of inspired air to expand and fill the airways. Expanding and filling airways takes energy and causes work—something we don't think about until it is usually too late (e.g., once smoking-impaired lung function is apparent). Therefore, in an otherwise healthy person, the dead space air partially reduces the work needed to reexpand the airways and to ventilate the alveoli during subsequent inspiration.

Inspired air is something that is on the minds of all experienced high altitude mountaineers. Reinhold Messner is one such mountaineer, adventurer, explorer, and author. He is renowned for being the first to summit Mount Everest without supplemental oxygen (most mountaineers of Messner's caliber climb using compressed supplemental oxygen). Messner is also famous for being the first climber to ascend all 14 peaks of 8,000-plus meters (26,000-plus feet above sea level; see, e.g., "Reinhold Messner," Wikipedia). During one of his record-setting climbs in 1978, Messner described periods of going in and out of consciousness and of seeing apparitions. In another 1970 climb, he lost six toes due to irreversible frostbite, and his younger brother Gunther, on the same climb, disappeared after an avalanche (see "Gunther Messner," Wikipedia; see also the movie *Nanga Parbat*, 2010).

In May 1996 and again in April 2015, tragedies occurred on Mount Everest. In the first, expedition leaders Scott Fischer (United States) and Rob Hall (New Zealand) and several of their adventure-seeking clients lost their lives when an unpredicted blizzard and human error divided and confused them in the "death zone" at 26,000 feet and above (see the movie *Everest*, 2015). Hall remained on or near the summit with a repeat client who was exhausted and refused to descend. Author Jon Krakauer was a member of the

tragic 1996 expedition and documented it in his best-selling book
Into Thin Air.

The April 25, 2015, earthquake in Nepal killed more than 9,000
people and injured another 23,000. It occurred at 11:56 a.m. local
time with a magnitude of 7.8. This was the worst natural disaster
to strike Nepal since the 1934 Nepal-Bihar earthquake. The 2015
quake triggered an avalanche on Mount Everest, where nearly
1,000 climbers were initiating the 2015 climbing season. The avalanche killed at least 19 climbers, making April 25, 2015, the dead-
liest day on the mountain in history. After a second quake on
May 12, 2015, all remaining climbers left. As a result, for the first
time in 41 years, no one summited Mount Everest in the spring of
2015. Avalanche-related disasters in 2014 caused the government
of Nepal to cancel that climbing season as well. Disaster-related
cancelation of back-to-back climbing seasons was another first for
Mount Everest.

Even with supplemental oxygen in one's inspired air, the challenge
of surviving in extreme environments like Mount Everest is multi-
pronged. Often by the time adventure seekers and paying clients and
their guides reach the summit (after an arduous midnight-to-early-
morning ascent), they are sleep deprived, malnourished, dehydrated,
hypoxic (supplemental oxygen supplies exhausted for some), hyper-
capnic, and otherwise physically, mentally, and emotionally spent.
In addition, some climbers simultaneously experience high altitude
pulmonary and/or cerebral edema (potentially life-threatening
mountain sicknesses) and might not even know it. Collectively, these
aphysiological states make climbers (even guides) prone to decisions
and mistakes they would not make in less harsh environments, thus
placing them and others at risk.

Spirometry and Respiratory Function Tests

The nasal, oral, laryngeal, and pharyngeal cavities, as well as the trachea and all airways down to and including the alveolar sacs in the average healthy adult, have a collective volume of about five to six liters. Over the years, respiratory physiologists have defined a set of volumes and capacities that are relatively easy to measure in the laboratory and that can be useful in the clinical diagnosis of respiratory disease and dysfunction. The branch of respiratory physiology that investigates these is called *spirometry*, and the apparatus used is called a *spirometer*. Today's spirometers are complex computerized devices that can be held in one's hand. However, such handheld electronic devices are not very useful in either teaching or learning about the important concepts of spirometry.

Classically, a spirometer consisted of two canisters, one with a larger circumference than the other. Both had one end opened to air. The larger canister was positioned upright and partially filled with water. The smaller canister was attached to a counterweight, chain, and pulley, then placed upside down and partially submerged in the larger canister. A breathing tube was passed through the water, emerging inside the inverted canister just above water level. The opposite end of the tube was placed in the mouth of the experiment's subject. When the subject exhaled, air entered the inverted canister and lifted it. When the subject inhaled, air left the canister and it sank deeper in the water. Respiration-related movements of the inverted canister could be collected on a recording device, and the volumes of air exhaled and inhaled could thus be estimated.

As mentioned above, the volume of air exhaled or inhaled by a subject during a single breath is called one's *tidal volume*. At the end of a normal expiration, any additional air that can be forcibly expelled from the lungs is called the *expiratory reserve volume*.

Conversely, at the end of a normal inspiration, any additional air that can forcibly be drawn into the lungs is called the *inspiratory reserve volume*. Air remaining in the lungs after even the most forceful exhalation is known as the *residual volume*. The total volume of air one can move in and out of their lungs in one minute is called the *maximal voluntary ventilation*; the maximum air one can move out of the completely filled lungs in the first second of a forcible expiration is called the *forced expiratory volume in one second*. Summation of two or more volumes is called a *capacity*. Examples of capacities include vital capacity, total lung capacity, and functional residual capacity.

Temperature and pressure, among other variables, affect the volumes of air described above. Air inside the lungs and respiratory conduits is saturated with the partial pressure of water vapor (humidity) and is usually at a temperature that differs from the atmospheric temperature (also known as *ambient* or *environmental temperature*). Therefore, corrections to volumes must be made according to these prevailing conditions. For such purposes, respiratory physiologists have created a set of guidelines to help investigators and clinicians. These include body temperature and pressure when inhaled air is saturated with water vapor (BTPS); ambient temperature and pressure when the air is saturated with water vapor (ATPS); and ambient temperature and pressure in arid conditions—that is, when there is little or no water vapor in the atmosphere and it is dry (ATPD).

In a healthy adult during quiet, resting respiration, tidal volume averages about 500 milliliters. On the typical Western diet (60–70 percent of calories coming from carbohydrates), our bodies consume more oxygen (approximately 250 ml/min) than they produce carbon dioxide (approximately 200 ml/min). This means that the volume of air inhaled during each respiratory cycle is slightly greater than the amount exhaled in the same cycle. In reporting changes in lung volumes, respiratory physiologists have chosen to measure the volume of air leaving the lungs rather than that entering them.

The magnitudes of inspiratory and expiratory reserve volumes depend on several things, including but not limited to one's (a) current lung volume, (b) lung compliance, (c) muscle strength, (d) comfort levels (e.g., pain tolerance limits), (e) musculoskeletal system flexibility, and (f) body posture. For example, the greater the lung volume after inspiration, the smaller the inspiratory reserve volume. A less compliant lung—that is, one that is more difficult to inflate—reduces inspiratory reserve volume. Muscle weakness or impaired innervation limits inspiratory reserve volume, and diseases of the skeleton (e.g., scoliosis, or curvature of the spine) and poor posture can affect inflation of the lungs as well.

Even after a maximal effort to eliminate it, a considerable volume of air remains inside the lungs. This is called the residual volume, and it amounts to two to three liters depending on the person and his or her age, gender, and health. Some have asked if having a residual volume reflects a flaw in the design of the respiratory system. The answer is a definitive *no!* In fact having a residual volume at the end of a normal, quiet expiration reveals the use of intelligence in respiratory design. First, the residual volume makes inflation of the lungs easier during the next inspiration. This is true because less energy and work are required to inflate a partially filled lung compared to an empty one. Remember, it takes more effort to inflate a completely deflated balloon than it does one that is only partially deflated. Try it. A second sign of intelligence in this design is the reserve oxygen the residual volume holds. Some of that oxygen is available for exchange with pulmonary blood should subsequent inspiration be impaired. If there were no residual volume and the lungs were completely empty after expiration, there could be no residual oxygen.

Hypoxia and Peripheral Chemoreceptors

Whether using supplemental oxygen or not, all climbers on Mount Everest and similar mountains are exposed to high altitude hypoxia. Fortunately, our cardiovascular and respiratory systems were designed with the above in mind. For example, there are sensory receptors located throughout the body that detect changes in chemicals inside the body. Two such sets of receptors are the central and peripheral chemoreceptors. They help regulate respiration. These receptors are strategically positioned to detect changes in oxygen and carbon dioxide in the blood and interstitial spaces. They are also wired to the brainstem and higher brain centers to be able to communicate with them. Since their discoveries, both sets of receptors have been extensively investigated by respiratory physiologists.

The peripheral chemoreceptors, called *carotid* and *aortic bodies*, are located at the branching of the common carotid arteries into external and internal branches. They are also located on the ventral surface of the aortic arch and in nearby vascular structures. The carotid bodies have been more thoroughly investigated than the aortic bodies because they are more accessible. Thus most of what we know about the peripheral chemoreceptors comes from research on the carotid bodies of animals and humans. Inside the carotid bodies are glomus cells. Glomus cells, the actual chemoreceptors, are not neurons but have several nerve-like characteristics including (1) voltage-dependent ion channels in their membranes, (2) innervation by autonomic sympathetic neurons for some glomus cells, (3) action potentials triggered by depolarization, and (4) intracellular, neurotransmitter-containing secretory vesicles. The most important function of peripheral chemoreceptors is to sense any lack of oxygen (hypoxia) in the systemic arterial blood and then to alert cells in the brainstem respiratory control centers to increase ventilation. Hypoxia stimulates, through sensory and motor nerve pathways, an increase in both the frequency and the depth of ventilation—that is, hypoxia stimulates an increase in tidal volume.

The first description of the function of peripheral chemoreceptors was published by Corneille Heymans. He was awarded the Nobel Prize in Physiology or Medicine in 1938 for his discovery. The sensitivity of the peripheral chemoreceptors is not limited to lack of oxygen. In the presence of normoxia (physiologically normal levels of oxygen in the blood), the carotid bodies are also sensitive to elevated concentrations of carbon dioxide (hypercapnia) and to reduced pH (acidosis). Both cases, like hypoxia, stimulate ventilation. Moreover, there are important interactions between the blood gases and pH in stimulating the glomus cells. For example, elevating the partial pressure of carbon dioxide augments the respiratory response to hypoxia; so does decreasing pH.

Besides their sensitivity to hypoxia, hypercapnia, and acidosis, other features of the carotid bodies are worth pointing out. In humans, carotid bodies are miniscule—each one weighing about two milligrams. Considering their size, they receive an unprecedentedly high blood flow that is greater than most other tissues in the body. When normalized for weight, blood flow to the carotid bodies is 40–50 times greater than that to the brain, one of the most well-perfused major organs in the body. In addition, carotid bodies have an unusually high metabolic rate—two to three times that of the brain. Thus they have an extremely high blood flow–to–metabolism ratio ensuring their role as chemoreceptors.

Sensory endings of the carotid sinus nerve (Hering's nerve) innervate the glomus cells of the carotid bodies. The carotid sinus nerve is a branch of the IX cranial nerve (the glossopharyngeal nerve). Sensory endings of the carotid sinus nerve pass into the brainstem and innervate respiratory control centers there. A decrease in oxygen or pH and/or an increase in carbon dioxide in the glomus cells cause them to release neurotransmitters. The neurotransmitters initiate action potentials in the carotid sinus nerve that travel to a respiratory control center (i.e., the dorsal respiratory group, or DRG) of the brainstem. The action potentials stimulate motor nerves that innervate the chest wall and other ventilatory structures. Collectively, these actions cause the rate and depth of respiration to increase, delivering more oxygen to the carotid bodies and exhausting excess carbon dioxide to the atmosphere. The physiological impediments (e.g., low oxygen) that initiated hyperventilation via this negative feedback loop are thus corrected, and respiration is restored to normal.

Imagine an exhausted mountaineer bivouacking overnight at 21,000 feet. During the night, her physical condition and high altitude hypoxia cause bouts of sleep apnea (cessation of respiration). As she stops breathing, the partial pressures and contents of oxygen in her lungs and blood decline. The ensuing hypoxia, accompanied by steadily rising contents and partial pressures of carbon dioxide in the blood, stimulate her peripheral chemoreceptors. Combined, these physiological activities excite respiratory control centers and stimulate breathing even before the climber is aware of her hypoxic condition.

Mechanistically, no one knows for sure how the peripheral chemoreceptors of the carotid bodies work. One hypothesis suggests that a heme-containing protein in the membrane of the glomus cell releases bound oxygen during hypoxia. Another hypothesis argues that the hypoxic state of the mitochondria causes an imbalance in the ratio of reducing and oxidizing agents in the cytoplasm of the glomus cell. A third idea regarding this molecular mechanism suggests an increased acidification of the cytoplasm during hypoxia. In each case, however, the glomus cell membrane depolarizes, causing calcium channels to open. Extracellular calcium diffuses into the cell and causes the release of neurotransmitters. Neurotransmitters create action potentials in the carotid sinus nerve, and respiration is augmented.

The strategic location of both carotid and aortic chemoreceptors also suggests an application of intelligence. Both locations are near the heart and not far from the brainstem. The carotid bodies in particular are nearly equidistant from the heart and respiratory control centers of the medulla. No matter how oxygen, carbon dioxide, and/or the pH levels of the blood change, the carotid bodies will be among the first tissues to experience this change. If these chemoreceptors had been randomly located in the knee joints or in the vascular beds of the toenails, another story of chemoreception and respiratory regulation would have to be told.

CARBON DIOXIDE AND CENTRAL CHEMORECEPTORS

Another set of chemoreceptors is located centrally in the medulla of the brainstem. These are called *central chemoreceptors*. When blood gases are at normal physiological levels, the central chemoreceptors are the primary source of negative feedback for evaluating ventilation. They are also the major source for tonic driving of the respiratory system. Central chemoreceptors are primarily sensitive to changes in the concentrations of hydrogen ions in the interstitial, extracellular fluids. Central chemoreceptors, like peripheral chemoreceptors, are neurons. However, instead of detecting arterial hypoxia (the job of peripheral chemoreceptors), central chemoreceptors detect arterial hypercapnia presented as respiratory acidosis. Respiratory acidosis is usually expressed as a decrease in the pH of arterial blood caused by a failure of the respiratory system to eliminate carbon dioxide. This carbon dioxide thus accumulates in the blood and leads to reduced pH.

Hydrogen ions, like other charged particles, cannot diffuse across the blood-brain barrier. Carbon dioxide, which is highly lipid soluble, diffuses across this barrier relatively easily, however. Once in the interstitial spaces that surround the central chemoreceptors, the excess carbon dioxide combines with water, producing other compounds including hydrogen ions. As long as the partial pressure of carbon dioxide (PCO_2) of arterial blood continues to increase or remains elevated, the level of hydrogen ions in the interstitial fluids will also be elevated.

In the 1950s, investigators used dogs whose carotid and aortic bodies had been denervated (i.e., there was no nerve communication between these chemoreceptors and the respiratory control centers of the brainstem). When the cerebral ventricles of these dogs were perfused with acidic solutions having an elevated PCO_2, the dogs hyperventilated. Because the resulting hyperventilation caused a corresponding respiratory alkalosis, the investigators reasoned that hyperventilation must have been caused by the acidity (resulting from elevated hydrogen ion concentrations and/or reduced pH) of the interstitial fluids surrounding the central chemoreceptors. From these and subsequent experiments, respiratory physiologists now believe that the primary stimulus driving hyperventilation during respiratory acidosis is not an increase in arterial PCO_2 but rather the ensuing decrease in pH within the surrounding brain tissues.

Beginning with an arterial PCO_2 that is physiologically normal (about 40 mmHg) and increasing this by about 12 percent, to 45 mmHg, can more than double ventilation (respiratory acidosis). Conversely, hypoxia doubles ventilation only when arterial partial pressure of oxygen (PO_2) is reduced by about 50 percent (i.e., from 100 to 50 mmHg). If the change in arterial carbon dioxide is accomplished suddenly, ventilation also increases suddenly. The change begins with an increase in the depth of ventilation, followed more slowly by a change in the frequency of inspiration. However, the response to respiratory acidosis might take 10 minutes or more to develop to the new steady state.

On the other hand, if the disturbance in acid-base homeostasis of the arterial blood is caused by a metabolic acidosis, the new steady state for hyperventilation occurs much more slowly, and the new steady state for ventilation is much less. Metabolic acidosis is achieved by causing the experiment's subject to breathe respiratory gases of a normal, fixed PCO_2 while reducing both the pH (elevating the hydrogen ion concentration) and the bicarbonate concentration of the arterial blood.

The reason for these experimental observations is that the central chemoreceptors, unlike the peripheral chemoreceptors, are located deep in the brain tissue outside of the cerebral blood vessels. They are bathed by the interstitial or extracellular fluids of the brain, which are separated from the arterial blood of the brain's cerebral arteries by a blood-brain barrier. This barrier is highly permeable to small neutral molecules such as oxygen (O_2) and carbon dioxide (CO_2), but it is almost impermeable to small charged ions such as hydrogen (H^+), sodium (Na^+), chloride (Cl^-), and bicarbonate (HCO_3^-). Thus an increase in cerebral arterial PCO_2 leads rapidly to a similar increase in PCO_2 in both the brain's extracellular fluids and the cerebrospinal fluid that bathes all brain tissues.

In actuality, it is rare for one arterial blood gas to change without causing corresponding changes in another. For example, it is uncommon for arterial PO_2 to decrease without causing accompanying increments in PCO_2 and hydrogen ion concentrations. Moreover, it has been found that an individual blood gas or pH can affect central and peripheral chemoreceptors differently. Thus it is of interest to respiratory physiologists and clinicians to know how the respiratory system responds to simultaneous changes in more than one blood gas parameter. Experimental respiratory acidosis (elevated PCO_2 at fixed and controlled pH, HCO_3^-, and PO_2) is informative in obtaining the answers.

When blood gases and pH are in their normal physiological states, respiratory acidosis has a greater stimulatory effect on respiration than does hypoxia. Usually, there is little carbon dioxide in the inspired air. When an experimental animal is ventilated on a mixture of gases containing carbon dioxide (e.g., 5–10 percent), the resulting respiratory acidosis causes ventilation to increase rapidly. Both central and peripheral chemoreceptors respond to such a challenge; therefore, either could be the cause of hyperventilation. Experimental physiologists have learned how to isolate the effects of the two populations of respiratory-related chemoreceptors. In one experimental approach, they denervate the peripheral chemoreceptors. This prevents the peripheral chemoreceptors from participating in the ventilatory response, thus isolating the contribution of central chemoreceptors. Conversely, to isolate contributions of the peripheral chemoreceptors, investigators developed perfusion techniques that allowed them to selectively produce changes in blood gases and pH at the peripheral chemoreceptors without disturbing corresponding chemistry at the central chemoreceptors.

They did this by interposing lengthy perfusion shunts between the carotid bodies and the central chemoreceptors. One end of the shunt was inserted into the internal carotid artery immediately downstream from the carotid bodies. The other end was inserted into the same artery even more distally (i.e., farther downstream). The volume of space in the shunt between the two points of vascular insertion ensured that it took flowing blood more than one minute, after it passed by the carotid bodies, to reach the central chemoreceptors. Before this time had elapsed, a respiratory response initiated by the peripheral chemoreceptors had already begun isolating the contribution of the peripheral chemoreceptors.

On the basis of such methodological approaches, it appears that the central chemoreceptors contribute up to 80 percent of the ventilatory response to respiratory acidosis during conditions of normoxia. However, responses of the peripheral chemoreceptors (seconds) are considerably more rapid than those of the central chemoreceptors (which can take several minutes to fully develop).

In the presence of hyperoxia, further raising alveolar PCO_2 has a linear effect on ventilation (respiratory acidosis, as stated above, stimulates ventilation). Repeating this experiment in the presence of hypoxia has a dual effect. First, reducing oxygen at a fixed carbon dioxide stimulates ventilation, demonstrating the effect of lack of oxygen on the peripheral chemoreceptors. Second, hypoxia increases the sensitivity of the central chemoreceptors to elevated PCO_2. This means that the slope of the curve reflecting the effects of graded respiratory acidosis on ventilation is increased in the presence of hypoxia. Investigators believe that at least part of the explanation for this interaction between hypoxia and hypercapnia is that the carotid sinus nerve itself becomes more sensitive to elevations in carbon dioxide during concomitant hypoxia.

Many people with lung disease have chronically elevated concentrations of carbon dioxide in the alveoli and in the arterial blood. One example of this is smoking-related emphysema, wherein destruction of alveoli and impairment of gas exchange between the alveoli and pulmonary capillary blood occur. Another example is with chronic neuromuscular and musculoskeletal conditions such as Lou Gehrig's disease (amyotrophic lateral sclerosis), wherein the ability to exhaust carbon dioxide to the atmosphere is impaired. Increases in the partial pressure of carbon dioxide in the lungs lead to an immediate accumulation of the gas in arterial blood (respiratory acidosis).

This will decrease pH in the blood and elsewhere. However, if the elevated partial pressure of carbon dioxide persists beyond four to eight hours, the pH in the interstitial spaces of the brain and the cerebrospinal fluid begins to improve.

GAS EXCHANGE AND TRANSPORT

My wife and I often take cross-country road trips. I love to drive and am increasingly annoyed and disappointed with the airlines. Since about 1971, we've crossed the United States in all directions multiple times. One of our constant companions on these drives, other than our children in earlier years, is the increasingly dense and dangerous tractor-trailer traffic. This traffic can be a nightmare on Routes 80 and 90 in and around the Chicago area and on Route 70 in the St. Louis area. Much of the country's east/west and north/south truck commerce converges on these two metropolitan areas. Motorists in much smaller and lighter automobiles can get trapped between and behind columns of trucks. This can cause these drivers to miss exit ramps, to act foolishly and hastily, and to make decisions they wouldn't make in less-tense circumstances.

Tractor-trailer traffic does not create a similar problem on the much more densely packed major highways of New Jersey (e.g., when one is traveling north or south between New York City and Philadelphia). New Jersey law forces truck and bus traffic to travel in lanes that are physically separated from automobile lanes on, for example, the New Jersey Turnpike (the world's most efficient superhighway) and the Garden State Parkway, where trucks and buses are not allowed. These laws result in less frustration and fewer accidents for all concerned.

It is common on our road trips to see full tractor-trailers traveling in any direction. It is not as common (but still evident) that many companies send trucks out full of goods but bring them home empty. This seems costly and inefficient and is arguably the cause of some trucking companies going out of business. Our intelligent bodies were not designed by the owners or operators of such businesses. The respiratory system is a good example of this truth. One of the main functions of the respiratory system is to deliver oxygen (cargo) from the alveoli (point A) to the tissues (point B). In this analogy, tractor-trailers on U.S. highways are red blood cells, and their cargo is oxygen and carbon dioxide. Carbon dioxide and oxygen are efficiently transported in combination with hemoglobin molecules inside red blood cells, or they are dissolved and carried as blood gases in solution (the water of blood plasma).

Home base for oxygen is the alveolus, where oxygen has a greater partial pressure than that arriving in the pulmonary arterial blood. Alveolar oxygen diffuses into the pulmonary capillaries and is then on its way to the peripheral tissues (point B, the distant customers in the tractor-trailer analogy). In the pulmonary venous blood, which after passing through the left atrium and left ventricle becomes systemic arterial blood, oxygen is either bound to hemoglobin or it is dissolved in the plasma. At the peripheral tissues, which have a lower partial pressure of oxygen than that of arterial blood, oxygen leaves the red blood cell (RBC) by diffusing down this partial pressure gradient and into the cells (point B, the customer's warehouse).

At this point, and if left to chance, the tractor-trailer (hemoglobin inside the RBC) could return to home base (an alveolus) with or without cargo. If empty, the transportation systems would be inefficient. Also, returning to home base without cargo (RBCs carrying carbon dioxide as they return to the alveoli) would make the respiratory exchange and transport system inefficient. Thankfully, the blood gas's exchange and transportation system is highly efficient (carrying loads in both directions), did not happen by chance, and will last a lifetime (approximately 120 years) if we take care of it and don't, for example, use tobacco or other drugs that damage the system.

The partial pressures and contents of carbon dioxide, being continuously produced by the actively respiring cells of the peripheral tissues (e.g., skeletal muscle, gut, heart, kidneys, etc.), drive carbon dioxide out of the cells and into the systemic capillaries. Once inside the systemic capillaries, carbon dioxide has two fates: one option is to remain in the blood plasma (11 percent), where it either dissolves (6 percent), binds to plasma proteins (less than 1 percent), or combines with plasma water to form carbonic acid, bicarbonate, and hydrogen ions (5 percent). Alternatively, it can enter red blood cells (89 percent), where it diverges into similar physiological states (e.g., it binds to hemoglobin that has just given up its oxygen to the cells, dissolves in intracellular water, or interacts with water molecules to become carbonic acid and then bicarbonate and hydrogen ions).

Our understanding of respiratory intelligence as represented by the exchange and transport of respiratory gases was greatly improved by Christian Harald Lauritz Peter Emil Bohr (Christian Bohr, 1855–1911) and John Scott Haldane (1860–1936). Christian Bohr, a Danish physician, discovered the influence that hydrogen ions and carbon dioxide have on the exchange and transport of oxygen. As arterial blood arrives at the peripheral tissues, its job is to unload oxygen and retrieve carbon dioxide. In this location, the cells and the tissues surrounding them have a lower pH and a higher carbon dioxide content than do the tissues of the lungs. This acidic environment facilitates the release of oxygen from hemoglobin and the subsequent binding of hydrogen ions and carbon dioxide to the newly freed hemoglobin molecule. Bohr discovered that both a low pH and an elevated carbon dioxide level greatly improve the ability of hemoglobin to release oxygen (the Bohr effect).

John Scott Haldane, a Scottish physiologist "famous for his intrepid self-experimentation" ("John Scott Haldane," Wikipedia) discovered the effects of oxygen on the ease with which the blood retrieves, transports, and releases carbon dioxide to the atmosphere. In environments where oxygen content is high, such as the alveoli of the lungs, hemoglobin gives up hydrogen ions and carbon dioxide much more easily than in environments where oxygen content is low. Once released from the pulmonary capillaries, carbon dioxide diffuses from an area with a high PCO_2 to an area of low PCO_2 (in this case, the alveoli). It is then released to an environment with an even lower content of carbon dioxide: the atmosphere.

Combined, the Bohr and Haldane effects are marvelous examples of the intelligence displayed by the respiratory system. The basic premise of both principles is that one process (e.g., oxygen delivery to the tissues) greatly facilitates the other as the cardio-respiratory systems try to maintain balance between whole-body oxygen supply and demand. Alternatively, the relationship between the Bohr and Haldane effects is the body's ability to intelligently balance production and removal of by-products such as carbon dioxide.

Ventilation Perfusion Ratios

Here *ventilation* refers to respiration and the exchange and transport of blood gases as described above. *Perfusion* refers to blood flow and the delivery of gases and nutrients to their needed locations. Both terms can be further defined by simple equations. Ventilation (V) is the product of respiratory frequency (RF), or the number of respiratory cycles per minute (cpm; i.e., the rate at which we breathe), and tidal volume (TV). TV is the volume of air flowing into the lungs during inspiration or out of the lungs during expiration, expressed in either milliliters or liters per breath. The simplest equation for this is ventilation = tidal volume × respiratory frequency, or V = TV × RF.

Perfusion (Q), on the other hand, is the product of heart rate (HR), or the number of cardiac cycles that occur in one minute (cpm), and stroke volume (SV). SV is the volume of blood ejected by either ventricle (left or right) during a single cardiac cycle. Stroke volume is also expressed in either milliliters or liters. The simplest equation for this is perfusion = heart rate × stroke volume, or Q = HR × SV. Perfusion, as used in this discussion and equation, is also called *cardiac output.*

The respiratory and cardiovascular systems are codependent. They interact continuously and were both intelligently designed to move loads (airflow and blood flow). Equally important, the volumes of flow they move must be matched in time and in space. It would make no sense if the respiratory system moved a volume of 5.0 liters per minute of air while the cardiovascular system moved a volume of 50 milliliters per minute of blood. Five liters (or 5,000 milliliters) is 100 times greater than 50 milliliters. Conversely, if the heart generated a cardiac output of 5,000 milliliters per minute while ventilation was only 50 milliliters per minute, there would also be no matching of the two quantities. Intelligently, the two systems were constructed so that each would produce a similar quantity of flow each minute. This makes perfect sense, since the function of the two systems is to support one another and the whole body by exchanging and delivering goods and services to the cells and by removing waste products from them.

We call this matching of ventilatory and circulatory functions the *ventilation-to-perfusion ratio* (V:Q). The ratio is generally about 1.0. That is, ventilation (airflow) and circulation (blood flow) both constitute quantities of about 5.0 liters per minute (5,000 milliliters per minute) in healthy human adults. Physiologists and clinicians split hairs and can show that the ratio is slightly less than or greater than 1.0 depending on body position and the region of the lung in which the measurement is made. For example, if one is lying down during sleeping hours, the ratio is 1.0 whether the measurement of airflow is made in the apex (top) or base (bottom) of the lung.

When one is standing erect, however, gravitational forces influence the distribution of blood flow to the lungs. Under these conditions, the V:Q ratio might be greater than 1.0 in the apex (creating more airflow and less blood flow) and as low as 0.8 in the base (more blood flow). Disease of the lung can also influence the ventilation to perfusion ratio. For example, the alveolar destruction caused by tobacco and other pulmonary toxins might be greater in a lobe of the apex than a lower lobe in the base of the lung. If this is the case, one can expect a marked decrease in the V:Q ratio in the apex primarily caused by reduced ventilation of that region.

Respiratory physiologists have spent considerable time investigating the matching/mismatching of airflow and blood flow to the lungs. They have developed equations and graphics that represent ventilation-to-perfusion ratios. Discussions of these advanced equations and graphics are usually limited to research and instruction in advanced graduate courses in respiratory physiology. They are beyond the scope of this book.

REGULATION OF VENTILATION

Aelius Galenus (Claudius Galenus, or Galen) was born in September 129 AD. His father, Aelius Nicon, was a wealthy architect and builder with interests in philosophy, mathematics, logic, astronomy, agriculture, and literature. Galen described his father as a just, good, and benevolent man. At that time, Pergamon (modern-day Bergama, Turkey) was a major cultural and intellectual center noted for its library (Eumenes II), which was second only to the Great Library of Alexandria, Egypt. About 145 AD, according to Galen, his father had a dream in which the god Aesculapius commanded Nicon to send his son to study medicine. Shortly thereafter, at age 16, and for the next four years, Galen began studies at the prestigious local sanctuary, or *asclepieum*, dedicated to Aesculapius, the god of medicine. Asclepiea functioned as spas or sanatoriums where the sick would seek the ministrations of the priesthood. Romans frequented the temple at Pergamon in search of medical relief from illness and disease.

In 148 AD, when Galen was 19, his father died, leaving Galen independently wealthy. Galen then followed the advice he found in the teachings of Hippocrates and began studying abroad, including in destinations such as Smyrna (now Izmir), Corinth, Crete, Cilicia (now Çukurova), Cyprus, and finally the Great Library and Medical School of Alexandria. In 157 AD, aged 28, he returned to Pergamon as physician to the gladiators of the high priest of Asia, one of the most influential and wealthy men in Asia.

Galen claimed that the high priest chose him over other physicians after he (Galen) eviscerated an ape and challenged other physicians to repair the damage. When they refused, Galen performed the surgery himself and won the favor of the High Priest of Asia. During his four years in Pergamon, Galen learned the importance of diet, fitness, hygiene, and preventive measures in caring for the body. Galen also learned living anatomy, and he viewed the treatment of fractures and severe trauma as windows into the body. Only 5 deaths of the high priest's gladiators occurred while Galen held the post, compared to 60 in his predecessor's time. The difference was ascribed to the attention Galen paid to their wounds (which probably included minimizing infections, a topic medicine would know little about for another 16 or 17 centuries).

Also during his treatment of wounded gladiators, Galen made a seminal observation. He noted a difference in respiratory and paralytic outcomes when combatants received wounds in the high versus the low cervical spinal cord. If the wounds were low (C5 or C6), the gladiator continued breathing but was paralyzed. If the wounds were high (C3 or C4), the gladiator stopped breathing. Galen repeated these observations in experimental animals and correctly concluded that the control of breathing begins in the central nervous system above the C3 or C4 vertebrae (i.e., in the medulla and pons of the brainstem). Thus began the continuing investigation of the role of the central nervous system in the regulation of respiration and ventilation.

Today we know there are groups of like-functioning neurons in these regions referred to as respiratory-related neurons (RRNs). They are found predominantly in the medulla of the brainstem in two locations, one called the *dorsal respiratory group* (DRG) and the other called the *ventral respiratory group* (VRG). We also know that both respiratory centers receive sensory input from multiple sources including higher brain structures (e.g., the hypothalamus and cerebral cortex) and the lungs and chest wall. Both sensory input and motor output in the respiratory centers get coordinated and integrated in the medulla by another set of regulatory neurons collectively referred to as the *central pattern generator* (CPG).

From the lungs, the conducting airways, and the expanding chest wall during inspiration, the vagus nerves (vagi) carry sensory information into the respiratory control centers. This vagal influence, coupled with other input, leads to cessation of inspiration and the onset of expiration. We know this is true because cutting the vagi bilaterally interferes with the termination of inspiration, thus prolonging it. When the vagi are cut, thus preventing sensory information from traveling into the respiratory control centers, there are also fewer respiratory cycles each minute. Moreover, each respiratory cycle is accompanied by a greater tidal volume. Thus the ventilatory apparatus produces sensory information via stretch receptors that detect expansion during inspiration. Under physiological conditions, this terminates inspiration and initiates expiration.

Neurons in the respiratory control centers are delicate and extremely sensitive to noxious stimuli. A wide variety of agents and actions can produce damaging, even life-threatening consequences for respiration. For example, narcotics such as heroin, opium, and morphine can depress and damage neurons in the respiratory control centers. Damage to these sensitive respiratory neurons is not limited to dangerous, illegal substances such as the opiates; it also includes the influences of alcohol, tobacco, other stimulants and depressants, and related compounds. Obviously, all such effects are dose-dependent and subject to other conditions such as the use of combinations of chemicals, the fitness and general health of the individual, and age.

It has been said by many that life begins and ends with breathing. Anyone who has observed the birth of a newborn or the death of someone is a witness to this statement. This author has observed both and is in agreement. After the creation of his physical body, the scriptures tell us that God then breathed into Adam his spirit—the breath of life—and Adam became a living soul. This implies an uncertain connection between breathing, the spirit, and the life of the soul. Anyone able and willing to reveal this mystery to me?

8

KIDNEYS AND
BODY WATER

My wife and I enjoy touring new model houses. She likes seeing floor plans and observing how different houses are designed. Most of the houses we tour have one, two, or three floors. Regardless of the number of floors and overall living space, there are kitchens, bathrooms, bedrooms, laundry rooms, and living rooms in each house. It goes without saying that each room has a different purpose and is occupied by different furnishings. For example, most kitchens are used for baking, cooking, and eating. Bedrooms are for sleeping.

Like model homes, our bodies have different rooms or spaces. We physiologists refer to them as *compartments*. For example, we call the space inside a cell the *intracellular compartment*. Spaces surrounding cells are known collectively as *extracellular compartments* or *interstitial spaces*. Blood circulates inside the vascular or intravascular compartment. Like rooms in a house, each of these separate physiological compartments contains elements or furnishings. Among other things, for example, the intracellular space has nuclei, mitochondria, ribosomes, and the endoplasmic reticulum. Such physiological furnishings are referred to as *subcellular* or *intracellular organelles*, and they each have designated functions.

Body compartments have at least two things in common. First, they contain water. Second, they have elements dissolved in water (e.g., electrolytes, ions, and inorganic and organic compounds). Electrolytes are elements like sodium (Na), chloride (Cl), potassium (K), calcium (Ca), and bicarbonate (HCO_3). These elements carry positive (Na^+, Ca^{2+}) and negative (Cl^-, HCO_3^-) charges, or they can be neutral and combined with other elements (e.g., sodium chloride, body salt, or common table salt). Electrolytes and ions are also distributed differently between the intracellular and extracellular compartments. In a state of physiological homeostasis, sodium is more abundant outside the cell and less abundant inside. Conversely, potassium is abundant inside the cell but is found in lower concentrations outside the cell. Generally speaking, the number of positively charged elements inside or outside a cell is balanced by a corresponding number of negatively charged elements. Also, the total number or concentration of electrolytes is the same inside and outside the cell.

Some elements have the ability to attract water. As such they are called *osmotically active*. Inside and outside the body, osmosis is what happens when water follows another chemical from one compartment or space to another. The combined osmotic activity of all elements inside or outside a cell is called its *osmolarity*, or *osmolality* (different units of expression). For the body to remain healthy and functional, the osmolarity must be the same in all three compartments at all times. In the homeostatic state, when all body compartments are in osmotic equilibrium, body osmolarity is about 280 milliosmoles per liter of water. This is the physiologically desirable state for body water and electrolytes.

Disturbances in osmolarity between different compartments can cause body spaces to swell or shrink, expand or contract. Swelling or shrinking of a compartment causes corresponding changes in the compartment's osmolarity. For example, if the extracellular compartment swells because too much water has been added, the corresponding osmolarity decreases to values lower than 280 milliosmoles. Such changes in volume and osmolarity, if sustained, can alter circulation and other cardiovascular hemodynamics. If changes in compartment size and osmolarity occur in the kidneys, they affect filtration, reabsorption, and the production of diluted or concentrated urine. Changes in osmolarity in the brain and lungs can affect cognition and the exchange of respiratory gases such as oxygen and carbon dioxide.

Among other things, the kidneys are tasked with regulating body water and body osmolarity. And the homeostasis of body fluids critically depends on the ability of the kidneys to decide how much of what chemicals to eliminate in the urine. The renal mechanisms that regulate elimination and retention of fluid, electrolytes, and other elements display important signs of intelligence. Consider a few thoughts of former international leaders on the topic. In the words of E. H. Starling (1866–1927), "The kidney presents in the highest degree the phenomenon of sensibility, the power of reacting to various stimuli in a direction that is appropriate for the survival of the organism; a power of adaptation that almost gives one the idea that its component parts must be endowed with intelligence." Or as H. W. Smith (1895–1962) put it, "Certainly, mental integrity is a sine quo non of the free and independent life. But let the composition of our internal environment suffer change, let our kidneys fail for even a short time to fulfill their task, and our mental integrity, or personality, is destroyed."

Osmoreceptors, Body Water, and Electrolytes

The sensory receptors that detect changes in osmolarity of body water are called *osmoreceptors*. Osmoreceptors are nerve cells (neurons).

They are located in the central nervous system above the brainstem but below the cerebral hemispheres, in a region near the third ventricle and hypothalamus. Osmoreceptors in this region respond to changes in osmolarity of the blood plasma and interstitial fluid. When osmolarity rises above or falls below 280 milliosmoles per liter, there is an imbalance. Consequently, distribution of water between the extra- and intracellular compartments changes, and the osmoreceptors either swell or shrink. In either case, the physical shape of the neuron changes. Changes in the osmoreceptors are transmitted to a second set of nearby neurons whose cell bodies are adjacent to the osmoreceptors. The second set of neurons synthesizes and transports the antidiuretic hormone ADH (also known as AVP, arginine vasopressin, named after an amino acid and the vascular actions of the compound) to the posterior pituitary gland. *Diuresis* means to eliminate excess water via the kidneys. *Antidiuresis* means to halt or diminish the flow of urine or loss of water from the body. The hormone is stored in and released from this location in response to needed adjustments to body water and osmolarity.

Another set of neuronal osmoreceptors is located near the set that stimulates release of ADH/AVP. Dehydration, increased osmolarity, and physical changes in the shape of these cell bodies stimulate thirst. Thus one set of osmoreceptors works to minimize the loss of body water by reducing urination. The second set of osmoreceptors encourages the direct consumption of water. This complements the reduction in urination. Jointly, the two sets of osmoreceptors, acting in coordination, correct the problem that activated them (i.e., increased osmolarity). Body water compartments are restored to a state of intelligent physiological homeostasis by changes in volume and osmolarity.

The complete cellular and molecular mechanisms for restoring body water and osmolarity are beyond the scope of this book, but a few details are worth knowing. Reduced excretion of water in the urine is accomplished by increased reabsorption of water by the more distal components of the renal tubules. From most to least distal, these tubules include (1) the outer medullary collecting ducts, (2) the inner medullary collecting ducts, (3) the cortical collecting tubules, and (4) the initial collecting tubules. These are all components of the individual nephron and are importantly involved in regulating body water and osmolarity. Osmoreceptors encourage consumption of water and renal tubules conserve water through reabsorption.

REGULATION OF BODY WATER AND SALT

Distinct but related control systems regulate the volume and osmolarity of the extracellular fluid. The body regulates the volume of extracellular fluid for many reasons including the control of blood pressure. If blood pressure is uncontrolled, then blood flow might also be uncontrolled and perfusion of the tissues will be at risk. Inadequate perfusion of organs leads to dysfunction (pathophysiology). Our bodies regulate the volume of extracellular fluid (including the vascular compartment) by regulating total body content of sodium chloride (osmolarity). This is important because hypo- or hyperosmolar states change the volume of water inside cells, which significantly affects cell function, especially in sensitive tissues like the central nervous system and lungs.

Volume of intracellular fluid is determined by the osmolarity of extracellular fluid. In turn, extracellular fluid osmolarity is determined by the concentration of sodium chloride in the extracellular fluid. This is a logical extension of these concepts: the body regulates osmolarity by regulating water content, and it regulates water content by regulating extracellular osmolarity. A stable water content means a stable osmolarity and vice versa. However, the two homeostatic mechanisms use different sensory receptors, different activators or effectors, and different hormonal/humoral integrators interposed between sensory receptors and effectors. Furthermore, both mechanisms have different effectors that are located within the kidneys. Thus our kidneys regulate extracellular fluid volume by controlling urinary excretion of sodium, and they regulate extracellular osmolarity by controlling the excretion of water.

After consuming a particularly salty meal, many people complain of swollen joints and tissues. This is especially true of those who dine in restaurants where cooks are known for adding extra table salt to most of the dishes. In worse case scenarios, one who dines out between 6–8 p.m. might not be able to form a fully clenched fist at 6–8 a.m. the next morning. His feet and ankles could also be swollen, and his shoes might not fit. The explanation for the unclenched fist and related symptoms is a temporary imbalance between body water, body salt, and their intra- and extracellular distribution. Edema of the digits and joints is the cause of the unclenched fist and tightly fitting shoes.

When one regularly consumes a diet high in salt, especially in the presence of renal dysfunction (e.g., known renal failure, diabetes), the body retains fluid. Both the volumes of plasma and interstitial fluid expand, causing swelling in more dependent regions such as the extremities. The swelling can be demonstrated by pressing an index finger against the swollen tissues (ankle, wrist, etc.). Depending on the severity of edema, an indentation and blanching (whitening of the skin) will occur. When the index finger is withdrawn, the indentation and blanching remain, a condition known clinically as *pitting edema*. This experiment can be performed on the same tissues when one has not consumed a diet high in salt and is experiencing no evident edema or swelling. Such experiences and experiments should encourage one to avoid frequent meals having high contents of salt.

Not all cases of pitting edema are caused by imbalances in salt and water. Peripheral vascular insufficiency, especially in the elderly, is another cause. This simply means that blood vessels, particularly veins (but not exclusively), are not functioning as they did in earlier years. If this is the cause of your swollen feet and ankles, then you should elevate your feet as often as possible. For example, lie on the carpeted floor with a blanket or other soft material beneath you. Stack two or three back cushions on the sitting cushions of your couch. Place a pillow beneath your head, get a good book, and extend your feet and legs onto the stacked cushions for 45–60 minutes. Repeat this several times each day, perhaps after meals and while watching the evening news.

The maintenance of water and salt balance also depends on physiological signals that reflect adequacy of the circulation. Adequacy of the circulation is determined by an effective circulating volume of blood. The effective circulating volume of blood is not an actual anatomical or measurable volume (such as plasma volume) but is rather the degree to which circulation meets the metabolic demands of the tissues. The effective circulating volume can be thought of as either an overfilled or underfilled vascular compartment. Special sensory receptors called *baroreceptors* apprise the central nervous system of the effective circulating volume. There are both low- and high-pressure baroreceptors. Intelligently, they have been strategically located within the circulatory system to detect changes in the effective circulating volume and to relay that information to the brain.

When the effective circulating volume is aphysiological, sensory afferent signals are conducted to the brain. There, integrating centers process the information and produce responses that correct extracellular fluid volume over short- and long-term periods of time. Short-term corrections occur over seconds to minutes. This is caused by autonomic nerves and other mechanisms acting to adjust functions of the heart and blood vessels. Long-term effects (hours to days) are mediated by mechanisms that adjust the renal excretion and balance of sodium. One is said to be in a physiological state of sodium balance when the intake of salt equals its output and has no untoward effects on the effective circulating volume.

HIGH-PRESSURE BARORECEPTORS

The main carotid arteries are located lateral to the trachea and buried between the muscles of the neck. Their pulsations can be felt by placing the combined index and middle fingers about 0.5–1.0 inches lateral to the trachea at the center of the neck. Of course, being able to find a pulse also depends on one's body morphology, body position, and weight. At the level of the lobes of the ears, each main carotid artery branches, forming internal and external carotid arteries. At these bifurcations, and within the first centimeter or so, the internal carotid artery appears swollen or enlarged. This region is called the *carotid sinus*. It is here that the main high-pressure cardiovascular baroreceptors are found.

High-pressure baroreceptors are sensory nerve endings that innervate the walls of the carotid sinuses. Baroreceptors were designed to detect stretch and de-stretch of the vessel wall throughout the cardiac cycle. During systole, when the heart is ejecting blood, the carotid sinus is stretched. In diastole, when the same volume of blood runs off to more peripheral vessels, the carotid sinus is de-stretched. Baroreceptors project sensory axons toward the brainstem in the form of a single carotid sinus nerve trunk. The carotid sinus nerve joins one of the main cranial nerves then passes into specialized regulatory regions of the brainstem.

From two lines of experimental evidence, we know that carotid sinus baroreceptors detect changes in pulsatile blood pressure. First, investigators have isolated the carotid sinus nerve and recorded action potentials passing over it during changes in pressure inside the carotid sinus. When blood pressure rises during systole and stretches the sinus, the number of action potentials increases. During diastole, when the sinus is de-stretched, the number of action potentials decreases. Second, when a stiff sleeve is snuggly placed around the carotid sinus to keep it from changing conformation during the cardiac cycle, the increased number of action potentials fired during systole in the absence of the sleeve is abolished. Many variations on these two experimental themes have been studied over the last two centuries. All results are consistent with the explanation just given.

The sensory information transmitted to the central nervous system by a stretching or de-stretching of the carotid sinus baroreceptor travels first to the medulla of the brainstem. In the medulla, there are cardiovascular, respiratory, renal, and other control centers. A control center is nothing more than a collection of nerve cell bodies performing similar functions. There are four different cardiovascular control centers. Two regulate the heart and two regulate blood vessels. The control centers are called *cardioaccelerator*, *cardioinhibitor*, *vasodepressor*, and *vasopressor* according to the actions they modulate. Activation or deactivation of a cardiovascular control center by an increased or decreased frequency of action potentials traveling to it causes one or more of the following actions: (1) an acceleration of heart rate, (2) a deceleration of heart rate, (3) relaxation of the walls of blood vessels and decline in blood pressure, and (4) contraction of the walls of blood vessels and an increase in blood pressure. Most commonly, a combination of responses is recorded and accounts for adjustments in heart rate, blood pressure, and the flow of blood.

As an example, consider that systolic and diastolic pressures in the carotid sinus are normally 120/80 mmHg (or less). These are physiological values, and the nerve activities traveling to and away from the cardiovascular control centers are in a steady state. Now imagine that a person was startled, administered a pressure-elevating drug, or contended with an office mate, and blood pressure is suddenly increased to 160/120 mmHg. An increased number of action potentials will be transmitted by the carotid sinus nerve to the medullary cardiovascular control centers. After interpretation and integration there, a set of efferent signals will travel from the brainstem to the heart and blood vessels, causing adjustments that restore blood pressure to 120/80 mmHg. A new postintervention steady state will be achieved, and cardiovascular homeostasis will redevelop.

RENAL BARORECEPTORS AND THE RENIN ANGIOTENSIN SYSTEM

The functional unit of the kidney is the nephron. In the human adult, there are about one million nephrons in each kidney. Nephrons are composed of tubules, blood vessels, and cuplike structures called *Bowman's capsules.* Inside a Bowman's capsule are collections of glomerular capillaries (glomeruli). Glomerular capillaries are the plasma-filtering units of the nephron. Leading to each glomerular capillary (glomerulus) is an upstream vessel called the *afferent arteriole.* Projecting away from the glomerulus is a downstream vessel called the *efferent arteriole.* Near Bowman's capsule, the adjacent afferent and efferent arterioles, and a segment of the thick ascending limb of the loop of Henle (near the beginning of the distal convoluted tubule) collectively form a specialized structure called the *juxtaglomerular apparatus* (JGA). The JGA is a fundamental communicating and regulating element of the kidney. Its three major components are the macula densa (MD), the extraglomerular mesangium (EGM), and the juxtaglomerular granular cells (JGA granular or secretory cells).

The macula densa is a plaque of epithelial cells found near the distal end of the thick ascending limb of the loop of Henle. Surrounded on all sides by epithelial cells of the thick ascending limb, the macula densa is consistently located a short distance, about 0.1–0.2 millimeters, from the end of the thick ascending limb of the loop of Henle, at its abrupt transition to the distal convoluted tubule. Because the nuclei of these cells are extremely large (compared to nuclei of epithelial cells in adjacent segments of the renal tubules), an unusually high nucleus-to-cytoplasm ratio causes the macula densa's relatively dense appearance, the distinguishing feature noted by early anatomists (ca. 1930s).

While mitochondria are numerous, they are not in contact with the epithelial cell membrane but are dispersed throughout the cell cytoplasm of the macula densa. The contact region between the glomerulus and the tubule is established early in the development of the nephron, but the embryologic cellular lineage of the macula densa is not known with certainty. However, the cell has some morphological similarities with the collecting duct, since both are cube-like and lack infoldings in their membranes. This suggests the two cell types might share a common cellular ancestry.

Early anatomists suggested that the nephron tubule appeared to be soldered to the vascular elements, and they described a polar cushion of cells joining tubule and blood vessels, now called the *extraglomerular mesangium*. Cells of the extraglomerular mesangium fill the wedge-shaped space between the cells of the macula densa and the glomerular afferent and efferent arterioles. This space lacks blood capillaries—a remarkable finding given the high density of microcirculatory vessels of the renal interstitium in general.

Nerve endings abound on the vascular elements and on the thick ascending limb in the region before the macula densa, but evidence strongly suggests that neither the macula densa nor cells of the juxtaglomerular mesangium receive direct innervation.

Abundant gap junctions connect cells of the extraglomerular mesangium with each other and with vascular elements in the juxtaglomerular apparatus. The electron microscopic evidence is consistent with the genetic expression of functional proteins called *connexins* in the juxtaglomerular apparatus. Conversely, no gap junctions or connexins have been found in cells of the macula densa, indicating that any tubular-vascular communication in this region probably employs diffusible chemicals such as paracrines or other elements of metabolism.

Juxtaglomerular granular cells have been described as cuboidal, epithelial-like cells in the media of glomerular arteriolar walls. These secretory cells are the main producers of the active aspartic acid protease renin. This is suggested by the fact that plasma renin concentrations decline to undetectable levels following removal of both kidneys. With a rough endoplasmic reticulum, a well-developed Golgi apparatus, and numerous cytoplasmic granules, granular cells of the juxtaglomerular apparatus have the fine subcellular structure of other protein-secreting cells. The renin-containing granules are membrane-bound and possess amorphous, electron-dense material believed to represent the mature form of the enzyme.

In a mature rat, kidney granular cells are clustered at the vascular pole over a length of about 0.03 millimeters, or about 20 percent of the length of the afferent arteriole. However, ringlike renin-positive regions in more proximal locations have also been reported. In the developing kidney and during stimulation of the synthesis of renin in the adult kidney, renin granules can be found in cells along the entire length of afferent arterioles and in larger upstream vessels as well. These observations suggest that vascular smooth muscle cells retain the potential for synthesizing renin throughout a lifetime.

In one method used to investigate the release of renin in rabbits, a short segment of the thick ascending limb of the loop of Henle (including the macula densa) and the corresponding juxtaglomerular secretory cells of the afferent arteriole are isolated from the rabbit kidney. The tubular segment of this preparation is then carefully perfused with solutions of known sodium chloride (NaCl) concentrations. Using controlled rates of flow, the downstream perfusate can be collected and analyzed for its contents of NaCl, renin, and any other chemical the investigators are equipped to measure. Such an approach permits the quantitative determination of the rates of release of renin as a function of the concentration of NaCl that the macula densa has been exposed to. The NaCl concentration at the macula densa is known and can be predictably altered. And all released renin can be collected and analyzed. Using this experimental approach, the renal baroreceptors and their sympathetic inputs can also be safely excluded.

While the above preparation is suitable for studying acute secretory responses of renin, it is not physiologically viable for periods that are sufficient to assess chronic changes in the release of renin or in the expression of renin-related genes. Still, the isolated perfused tubule methodology has shown definitively that increasing NaCl concentration in the tubular fluid at the macula densa inhibits release of renin and that reducing NaCl concentration stimulates it. It is noteworthy that the renal baroreceptor mechanism has also been studied using perfused afferent arterioles (juxtaglomerular secretory cells) rather than perfused renal tubules. Relative to the release of renin, results with this approach are consistent with those obtained in perfused tubules. Such experimental findings are consistent with the kidney's intelligent ability to regulate renal perfusion pressure, glomerular filtration rate, and the retention or elimination of sodium to maintain physiological harmony within the body.

In the late 19th century, investigators observed that injection of crude extracts from the kidneys caused a slow-developing but sustained increase in systemic arterial blood pressure. Those physiologists called the active ingredient in the extract *renin*. It was later shown that repetitive injections of crude renin caused successively weakening vasopressor responses. This property was given the name *tachyphylaxis*. Later still, it was determined that purification of the crude protein enhanced its abilities to elevate blood pressure. However, when purified renin was injected into a rabbit's ear vein that was perfused with Ringer-Locke solution, the potency of the renin was reduced. Ringer-Locke, like several other solutions, are physiological salt solutions; their chemical composition is designed to replicate that of blood plasma. When rabbit plasma was added to Ringer-Locke solutions, renin caused intense constriction of the rabbit's blood vessels. Collectively, these results suggested that another substance, found only in plasma, was necessary for renin's pressure-elevating effects.

In the 1930s and beyond, experimental physiologists found that partial occlusion of a renal artery caused elevations in blood pressure (hypertension). For example, partial occlusion of one renal artery in a dog led to elevated blood pressure that returned to normal only after several weeks. If both renal arteries were partially occluded, or if one artery was occluded and the opposite kidney removed, the hypertension became persistent and stable. Additionally, in rats, rabbits, goats, and sheep, constriction of one renal artery led to elevated and persistent changes in blood pressure. If constriction of the artery was excessive, pronounced hypertension, renal failure, and death ensued.

At about this same time, investigators in the United States and Argentina reported that renin is a proteolytic enzyme (i.e., it can break down proteins). They hypothesized that renin reacts with a chemical in the plasma to produce a nonprotein vasopressor substance each team, respectively, called *angiotonin* and *hypertensin*. In the interest of scientific goodwill within the Americas, both groups agreed to call the new substance *angiotensin*. Thus the renin-angiotensin system was introduced to the world of physiology and medicine.

Now nearly 90 years later, we know that the plasma substrate upon which renin acts is a large, globular protein produced by the liver. We call it *renin substrate*, or *angiotensinogen*. When renin comes in contact with this macromolecule, it snips a bond between two adjacent amino acids, creating a physiologically inert, much smaller protein containing only 10 amino acids and called *angiotensin I* (AI). As AI passes through the pulmonary microcirculation, it comes in contact with an enzyme called *angiotensin converting enzyme* (ACE). ACE is found primarily on the endothelial lining of pulmonary capillaries. ACE clips off two amino acids from AI, creating a powerful vasopressor (vasoconstrictor) substance called *angiotensin II* (AII).

In experimental animals and in humans, intravenous injections of small amounts of AII cause sharp and pronounced increments in blood pressure that last several minutes. Conversely, when purified renin is injected, pressure rises slowly but can be sustained for more than an hour. The prolonged action of renin results from the slow but continuous release of AII from the renin substrate. A third enzyme, angiotensinase, is also present in circulating plasma and comes from tissues such as the liver, gut, and kidneys. It degrades AII and accounts for termination of its vasopressor effects.

What happens if blood pressure in the renal artery decreases—that is, what if renal perfusion is impaired by partial occlusion of a renal artery? Decades of research provide convincing evidence that when this happens, there is an interaction between degranulation of the secretory cells of the polar cuff of the afferent arteriole, activation of the renin-angiotensin system, and stimulation of renal sympathetic nerves. The polar cuff is both a pressure-sensing and a hormone-secreting organ, and no fewer than four factors control the rate of release of renin from the polar cuff. They are (1) a change in the intraluminal pressure in the afferent arteriole leading to the cuff, (2) a change in the frequency of sympathetic efferent action potentials reaching the polar cuff, (3) endogenous catecholamines and other neurohumoral elements circulating to the JGA, and (4) a change in the chemical composition of the tubular fluid reaching the macula densa.

After the earliest demonstrations that constriction of a renal artery elevates systemic arterial blood pressure, investigators began wondering if renal ischemia was the causative mechanism for release of renin. Subsequently, both renal hypoxia and renal hypercapnia were ruled out as effective stimuli for renin's release. Nor is a reduction in renal pulsatile pressure an effective stimulus even though a modest decrease in renal mean arterial pressure (one so low that renal blood flow is not effected) does cause significant release of renin. The theory has evolved, therefore, that the release of renin from the JGA is exquisitely sensitive to reductions in mean arterial pressure inside the afferent arteriole.

STARLING FORCES AND THE RATE OF GLOMERULAR FILTRATION

A physiological renal arterial perfusion pressure is needed to maximize the plasma filtration process in the glomerulus. When this happens, glomerular filtration rate (GFR) in the normally healthy young adult human is about 120 milliliters per minute, or 180 liters per day. This rate of filtration ensures multiple exposures of the 2–3 liters of circulating plasma to the combined glomerular filters of both kidneys each day. Such filtration prevents accumulation of waste products of metabolism as well as by-products of prescription (and other) medications. It also helps maintain the acid-base balance. These things cannot happen with sustained elevations or reductions in GFR. Moreover, because GFR routinely decreases with age as the mass of renal tissue and glomerular filters decreases, the elderly are at increased risk of the toxic accumulation of by-products from the dispensary of medications they have been prescribed.

Ernest Starling and his colleagues were the first to explain the role of capillaries and the microcirculation in maintaining circulating blood volume and other hemodynamics connected to it. Starling demonstrated that four forces are at play inside and outside every capillary. Using glomerular capillaries as an example, these forces are (1) glomerular capillary hydrostatic pressure, (2) glomerular capillary oncotic pressure, (3) hydrostatic pressure inside Bowman's capsule, and (4) oncotic pressure of the fluid inside Bowman's capsule. Constancy in the numeric values of these four physiological factors (Starling forces) is necessary to yield a GFR of 120 milliliters per minute, or 180 liters per day.

Two of the above Starling forces favor movement of plasma water and its solutes from inside the glomerular capillaries to the outside (into Bowman's space). These forces are the glomerular capillary hydrostatic pressure and oncotic pressure inside Bowman's space. The remaining two forces—glomerular capillary oncotic pressure and hydrostatic pressure inside Bowman's space—oppose the movement of water and solutes from the glomerular capillaries into Bowman's space. Alternatively, these remaining two forces favor uptake of water from Bowman's space into glomerular capillaries. To be in a state of physiological harmony, the four forces must be balanced at all times.

Oncotic pressure, also called *colloid osmotic pressure* (or *colloid oncotic pressure*, among other terms), is caused by the presence of proteins inside capillaries. Proteins bind smaller charged inorganic elements, and these charged elements are often hydrated (i.e., they attract water). Additionally, the amino acids from which proteins are constructed are charged themselves (except for the neutral amino acids) and can attract charged, hydrated inorganic elements such as Cl^- and Na^+. Where ever two compartments of fluid are separated by a selectively permeable membrane, and where charged proteins exist on only one side of the membrane, water will be attracted to that location (hence the designations *colloid osmotic* or *colloid oncotic pressure*). The two fluid compartments and the selectively permeable membrane being referred to here are, respectively, the space inside the glomerular capillaries (intravascular compartment), the space inside Bowman's capsule (extravascular compartment), and the glomerular capillary wall (the barrier of endothelial cells and other elements that form the selectively permeable membrane).

Mathematical equations developed by Starling and others as well as direct measurements have allowed physiologists to assign values to each of the four Starling forces. This has been done in many capillary networks—not just in glomerular capillaries and Bowman's spaces. For example, under the lens of a light microscope, investigators have impaled the capillaries at the base of their fingernails using glass pipettes with tips drawn to an internal diameter of fewer than six micrometers. By adjusting the hydrostatic pressure inside the glass pipettes and by watching movements of red blood cells, they have been able to estimate hydrostatic pressures inside the capillaries. Modifications of this technique have been used to measure hydrostatic pressure in the capillaries of other organs—for example, in mesenteric capillaries in the connective tissue that holds the loops of gut together, capillaries of the bat wing and rabbit ear, and so on. A. C. Guyton and his colleagues at the University of Mississippi in Jackson were among the first to attempt the direct measurement of hydrostatic pressure outside the capillaries, in the interstitial spaces.

When the algebraic sum of glomerular capillary hydrostatic pressure and oncotic pressure in Bowman's space is greater than the corresponding sum of glomerular capillary oncotic pressure and hydrostatic pressure in Bowman's space, net filtration of water and solutes will take place. That is, water and solutes will leave the glomerular capillaries and form a volume of filtered fluid inside Bowman's space called an *ultrafiltrate* (because it resembles blood plasma without the proteins). Under physiological conditions and in any period of time, the vast majority of filtered water and solutes (greater than 99 percent) gets reabsorbed by the nephron tubules. The ultrafiltrate that is not reabsorbed is excreted as urine.

DISRUPTION AND CORRECTION OF WATER BALANCE

There is a regulated balance between the volume of water consumed in a day and the volume excreted by the kidneys. The same can be said of the mass of osmotically active particles consumed and excreted each day—for example, salt. There are three different primary modes by which water can enter our bodies (water consumption): direct, indirect, and intrinsic. A fourth mode is via mitochondrial metabolism. In the latter process, and in the presence of oxygen, carbohydrates, fats, and proteins are converted into energy, water, and carbon dioxide. We consume water directly by drinking it from the tap or from commercially-bottled sources. Indirect water consumption comes in the self-made or commercially available drinks we consume, such as coffee, sodas, tea, ad infinitum. Intrinsic water is found in fruits and vegetables, although most foods contain various amounts of water.

As I have stated elsewhere, the kidneys are charged with regulating body water and its distribution, even though water is lost from the body by routes other than the kidneys (e.g., in the feces, through respiration, and by sweating). The volume of water lost through sweating can vary greatly depending on environmental temperatures and physical activity. However, loss of water through sweating serves the purpose of regulating core body temperature rather than the volume and distribution of body water. Kidneys adjust the excretion of water to compensate for variability in the volumes of water humans consume.

Obligatory loss of water is a key principle in the kidney's regulation of body water and salt. *Obligatory* means that no matter what the body's contents of salt and water are, the kidneys are obligated to excrete sufficient water to dissolve any solutes that are simultaneously excreted in the urine. Regardless of the volume of water they eliminate, the kidneys must excrete 600 milliosmoles of solute each day. On the average Western diet and with physiological consumptions of water and food, the kidneys excrete these 600 milliosmoles of solute in 1,500 milliliters of water each day. In other words, the product of the volume of urine eliminated per day and the osmolality of the urine are nearly constant. It follows that in order for the kidneys to eliminate a wide range of volumes of water, they must also excrete a wide range of milliosmoles in the urine. That is to say, urine can range from very diluted to markedly concentrated as far as solutes are concerned.

When a healthy young adult is excessively hydrated, the kidneys can eliminate as many as 20 liters of urine per day. Under these conditions, instead of the urine having a normal osmolarity of about 400 milliosmoles per liter of urine, it can have as few as 30 milliosmoles per liter. This would be considered a very diluted urine. Conversely, when the body is superdehydrated, there is little urinary water to eliminate, because the kidneys were designed to conserve water during dehydration. When superdehydrated, urine can have an osmolarity of 1,200 milliosmoles or more per liter, or about three times the normal level. Thus one can see that the kidneys are more suited to eliminating water (producing diluted urine) than to conserving it (producing concentrated urine). Still, the mechanisms regulating both water diuresis and antidiuresis reveal signs of renal intelligence at every step.

What actually happens when the intake of water is severely restricted and the kidneys become responsible for conserving body water (antidiuresis)? For certain, one gets thirsty. This drives the individual to consume water if it is available. We also urinate less frequently. Dehydration and reduced urination result in a greater concentration of milliosmoles in the volume of urine that is excreted. Additionally, concentrated urine has a darker, more definitive yellow hue than diluted urine. This means that the concentration of urine-coloring chemicals such as the hemoglobin by-products bilirubin and biliverdin is increasing.

If water restriction is severe, the only urination we do is to eliminate milliosmoles and by-products of metabolism (creatinine, uric acid, hydrogen ions, etc.). This is what obligatory water loss means. Under such conditions, the volume of water excreted by the kidneys will be only the amount needed to dissolve the solutes in the urine. For example, instead of eliminating 1,500 milliliters of water per day under normal conditions of hydration, the kidneys might eliminate only 150 milliliters per day in cases of severe restriction of water.

LOW-PRESSURE BARORECEPTORS

The baroreceptors located in high-pressure sites of the circulatory system (e.g., the carotid sinus and the aortic arch) are not the only stretch receptors that provide feedback regulation to the circulatory system and kidneys. Another set of baroreceptors is found in the walls of the atria, the right ventricle, and in the nearby vasculature (e.g., the pulmonary artery, the pulmonary veins, and the inferior and superior vena cava). These are low-pressure mechanoreceptors. They are the bare ends of the myelinated nerve fibers that innervate these important cardiovascular structures. Stretching of these structures depends primarily on the volume of venous blood returning to the left and right sides of the heart. These mechanoreceptors detect the so-called fullness of the circulation. They are part of a larger system of sensory receptors that help regulate the effective circulating volume of blood.

Atrial baroreceptors are the most extensively studied of the low-pressure mechanoreceptors. They are located at the ends of afferent sensory axons known as A and B nerve fibers. The A fibers generate action potentials during atrial systole (atrial contraction) and can therefore be said to monitor heart rate (since firing of the pacemaker of the heart, the SA node, is what triggers atrial systole). The B nerve fibers fire during ventricular systole. They steadily increase their action potential firing rate as the ventricles fill, reaching maximum frequency at the peak of the atrial or jugular pulse. B fiber baroreceptors monitor atrial and ventricular filling. Central venous pressure—that is, the pressure inside large veins leading to the right side of the heart—is the main variable that determines right atrial filling. Thus B fibers also monitor changes in central venous pressure. An extension of this argument is that B fibers, by inference, also monitor effective circulating volume.

The afferent pathways for the low-pressure baroreceptors are the same as those for high-pressure baroreceptors. They join the tenth cranial nerve (the vagus nerve) and travel to the nucleus tractus solitarius and other nuclei of the brainstem medullary cardiovascular control centers. To a considerable extent, the efferent pathways for the low-pressure mechanoreceptors and the high-pressure receptors are the same (heart and blood vessels are the effector organs). Stretch of the high-pressure baroreceptors causes bradycardia, or a reduction in heart rate (the baroreceptor reflex), whereas stretch of the low-pressure baroreceptors causes tachycardia, or an increase in heart rate (the Bainbridge reflex).

Also, increased stretch of the high-pressure baroreceptors causes generalized systemic vasodilation, whereas increased stretch of the low-pressure atrial B receptors causes only renal vasodilation. The net effect of increased atrial stretch (tachycardia and renal vasodilation) is an increase in renal blood flow, glomerular filtration rate, and renal excretion or urine output (renal diuresis). Decreased stretch of the low-pressure atrial B receptors has little effect on heart rate but increases sympathetic nerve activity to the kidneys, resulting in reduced renal blood flow, glomerular filtration rate, and urine output (renal antidiuresis). High-pressure baroreceptors respond to stretch (elevated blood pressure) by trying to reduce blood pressure. Low-pressure baroreceptors respond to stretch by trying to eliminate fluid.

In both cases, the response of baroreceptors to increased stretch is an intelligent one. Elevated blood pressure damages the vasculature and can harm the organs and tissues it serves (e.g., high pressure-induced damage to the retina of the eye in those with diabetes). Overstretching of low-pressure baroreceptors promotes diuresis (loss of water) but can also promote natriuresis (loss of sodium). The atrial myocytes themselves, when overstretched, release a substance called *atrial natriuretic peptide*. This agent causes the kidneys to excrete sodium. Renal excretion of sodium and water can be helpful to distressed men whose urinary bladders are compressed by enlarged prostates.

THE URINATION REFLEX (*MICTURITION*)

Micturition and urination are the same thing. Urination is controlled by a complex process called the *micturition reflex*. It involves the body, neck, and sphincters of the urinary bladder, systemic and autonomic afferent and efferent nerve tracts, a control center in the brainstem (the micturition center), and higher central structures in the midbrain and cerebral hemispheres. In a healthy young man or woman ages 20–30 years, the first clue of a pending need to urinate occurs at a bladder volume of about 150 milliliters. When the bladder stores 300–400 milliliters of urine, a much stronger sensory desire to urinate occurs. At a maximum bladder capacity of 500–600 milliliters, the person is markedly uncomfortable, and the need to urinate is beyond urgent. Irreversible, acute damage to upstream structures, including the kidneys, can occur if the bladder is not emptied.

Bladder function can be investigated in the office of a nephrologist or urologist. The patient is asked to empty their bladder, and a sterile catheter is passed through the urethra and into the bladder. Any residual urine in the bladder is drained through the catheter. The catheter is a double-lumen device. One lumen is used to fill and drain the bladder. The other lumen contains a pressure sensor for monitoring pressure inside the bladder. Once the bladder is emptied and a resting, baseline pressure is recorded, the physician incrementally refills the bladder with sterile, warm water or a warm saline solution. Fifty milliliters of water or saline are injected into the empty bladder, and a few minutes later pressure is recorded. This is repeated multiple times until the bladder is filled maximally or until the patient urinates.

After injecting the first 50–100 milliliters of solution, pressure inside the bladder rises sharply. This is because the bladder is not very compliant when it is empty. As the bladder continues to fill to 100–400 milliliters, the pressure-volume relationship is not as steep—a plateau is achieved. During this period of filling, the bladder wall is much more compliant due to active relaxation of the smooth muscle that comprises the body of the bladder. Then between 400–600 milliliters or so, there is another sharp rise in pressure as volume continues to increase. At this point, if the patient does not willfully override the urge to urinate, the micturition reflex is activated, and the bladder begins to contract intermittently until it has emptied.

The diagrammatic representation of the relationship between pressure and volume inside the bladder is called a *cystometrogram*. That information and other characteristics of the patient's patterns of urination, as well as their age and general health, can help the clinician diagnose malfunctions of the bladder such as overactive bladder (OAB) in women and benign prostatic hyperplasia (BPH) in men.

Details of the neurophysiology of the micturition reflex are beyond the scope of this book. However, being familiar with the patterns of innervation of the bladder, its sphincters, and surrounding structures can be helpful, especially to the patient considering options for medical or surgical treatment of OAB or BPH. It is a virtual impossibility, no matter how experienced the urologist or robot, to surgically correct a dysfunctional bladder without damaging nerves, blood vessels, and other tissues. Moreover, no medical treatment—whether it be radiation, chemotherapy, or other drugs—is without toxic and potentially life-threatening side effects.

In about 2009, I was diagnosed with an enlarged prostate (BPH). The diagnosis was based primarily on physicians' examinations and steadily rising plasma concentrations of prostate specific antigen (PSA). Although widely debated by medicine and insurance companies today, PSA concentrations remain the standard for the initial diagnosis of BPH. Prior to such diagnoses, PSA concentrations for unaffected men range from 0 to 1 nanograms per milliliter of plasma. Over a five- to six-year period, my PSA concentrations rose from less than 1.0 to over 7.0 nanograms per milliliter. The urologist I was seeing advised a prostate biopsy and a second opinion. I found another urologist who had earned both PhD and MD degrees. This suggested that he probably had a better appreciation for research and experimentation, so I chose him.

At our first appointment, I mentioned my background in physiological research. I asked if he would be willing to pursue an experimental approach with me, and he agreed. It was already known that drugs used to treat other male-related renal and reproductive dysfunctions (e.g., Cialis, Levitra, and Viagra) have been helpful in treating BPH in some men. We discussed this possibility, and my urologist recommended a 30-day trial of Cialis before he collected biopsies. I agreed. The recommended dose of Cialis for BPH was 5 milligrams each day for 30 days. I started taking the drug on a Wednesday. By the following Friday, I was experiencing disturbing side effects. These intensified on Saturday, after just 4 days and a cumulative dose of 20 milligrams.

The side effects were expressed in my musculoskeletal system and occurred from my knees to my shoulders and neck. They were isolated to the posterior half of my body, and I felt rheumatism-like pain while sitting and standing. The symptoms intensified when I was lying down in the supine position, and I could not fall asleep as easily. I reasoned that if the pain was caused by Cialis and not some

other event, the symptoms should disappear quickly if I discontinued the drug, which I did. After two days (Sunday and Monday), my
symptoms were gone. I concluded that five milligrams of Cialis daily
was not good for me and that the drug caused neuromuscular toxic-
ity as well as generalized impairment of my musculoskeletal system.
I never resumed taking Cialis.

When this experimental urologist performed his digital examination of my prostate, he concluded that it was about 40 grams in
mass. He said it was definitely enlarged, but based on his experience
with men of my age and body morphology, my gland was not unusually large. This gave me some comfort. However, we did agree that I
should have an MRI and biopsies shortly thereafter. According to the
radiologist's report, the MRI images allowed him to estimate a size of
about 40–50 grams (enlarged but not unusually large). The biopsies
suggested that the enlargement was not cancerous and the prostate
did not appear to be metastasizing. Based on these results, we agreed
that I should do nothing except wait watchfully until the next MRI or
biopsies. The urologist suggested waiting another five years. Because
I had already learned how to live with my symptoms, I could continue doing this for the next several years.

As Ernest Starling hinted more than a century ago, a careful
observer might conclude that the kidneys possess intelligence. I
have taught renal physiology to advanced undergraduate students
for many years. I have also experimented with cardiovascular variables that profoundly affect renal function. My experience con-
vinces me that Starling was right—the kidneys do display multiple
signs of being endowed with intelligence.

9

GUT AND NUTRIENT FLOW

No two physicians, patients, or general citizens define fasting the same way. Speaking physiologically, I fast every 24 hours. So do you if you are one of the average adults in society. After my evening meal, if I don't consume food again until breakfast, I have fasted, arguably, for about 12 hours. I wake up hungry, and my circulating concentrations of blood sugar (glucose) are at their lowest daily levels. We call our morning meal *breakfast* because the word actually means "breaking a fast." Thus this is a defendable definition of a period of physiological fasting.

There is neither time nor space here to discuss all the physiological and biochemical events that occur between 6 p.m. and 6 a.m. that help define the daily physiological fast. However, these events are thoroughly documented because they have been under investigation by physiologists, nutritionists, dieticians, and clinicians for more than a century. I encourage the interested reader to do a little investigating themselves.

There is also the "starvation" fast. Some prisoners and activists engage in this kind of hunger strike. I do as well on occasion. For example, during the winter holidays, I can easily overeat and gain five or more pounds. I am not sufficiently patient to rid myself of this unwanted baggage over an extended period of time. So to eliminate the extra weight, I might skip a meal here, consume fewer calories in several meals there, and wait longer periods of time between meals. Using these activities, I can confidently lose the extra five pounds in a few days. I have done my best over the decades to not allow these extra pounds to accumulate. As a consequence, I still weigh the same today as I did when leaving basic training in the army in 1967.

Historically, one of the best discussions for why we should fast was given by Isaiah and is recorded in the Old Testament. Chapter 58 of Isaiah is wonderful reading for at least two reasons. First, Isaiah helps us choose some life-altering purposes for our fasts (Isaiah 58:6, 7 KJV). Second, he gives a list of personal blessings available to any person who fasts for the right reasons (Isaiah 58:8–14). One of the promises Isaiah makes is that one's health can spring forth speedily if the person fasting follows Isaiah's instructions. He also tells us that during periods of drought, the successful faster will be like a watered garden.

I love the imagery Isaiah's writings evoke. Who would not like to have their health spring forth speedily and to live in a physiological state akin to that of a well-watered garden, metaphorically speaking? My wife has several gardens. When they are watered and at their biological peaks (late May / early June), they beautify and enhance the appearance of our home and property. They also attract the gazes and compliments of passersby. Most importantly, they help feed us. One of the reasons I still push a lawn mower is not only to take time to observe her flowers, grasses, shrubs, and trees but also to regularly pick a fresh, warm tomato to eat while mowing the lawn.

As I have already mentioned, fasting can be taken to extremes, and under such conditions, it serves little or no useful purpose. Conversely, the physiological and medical benefits of purposeful, reasonable fasting are too many to document here. However, a few are worth noting. There is good experimental evidence that fasting activates lipolytic enzymes, which help metabolize fat (lipids). Activation of such enzymes mobilizes stored fat and reduces its content in cells. This provides energy when it is not otherwise available and helps with loss of unwanted weight. Fasting also mobilizes glycogen (a form of stored glucose) in the liver, thus sustaining circulating concentrations of glucose and other energy sources. Moreover, fasting gives the pancreas a break from the constant demand eating places on it to produce, store, and release insulin and other hormones.

I will have more to say about the physiological differences between the fed and fasted states in later sections. For now, suffice it to say that fasting is an important practice serving both physiological and nonphysiological purposes. Fasting regularly (e.g., monthly) is to the body what crop rotation is to a field. Not planting a crop for six or seven years gives the field a chance to rest. During the season of rest, the soil is reaerated and rewatered without demand. Nutrient stores can accumulate. And fertilizer-related waste products can be washed out or otherwise diluted before another planting season begins. Fasting gives our secretory cells a rest and should help us fill our minds while ignoring our empty stomachs.

THE GASTROINTESTINAL SYSTEM AND SECRETION

The physiological functions of our gastrointestinal systems can be grouped into the broad areas of secretions, mobility and mechanics, and digestion and absorption. There are secretory cells—pits and glands distributed throughout the system from oral to aboral (anal) ends. Decades of research have documented the effects of the sounds, sights, and aromas of meals being prepared on our secretory function. We have all heard someone make a statement such as, "The aroma of that dish just makes me salivate." It is true that acting alone or in combination, the sights, smells, and sounds of food being prepared in a kitchen stir our appetites, especially when we are hungry. Such events prepare us for the subsequent consumption of a meal. The preparations begin in the mouth but certainly do not end there.

Long before we take the first mouthful of a meal, our salivary glands have been at work producing secretory products. These products include buffers, electrolytes, and enzymes that are released into the oral, laryngeal, and pharyngeal spaces. For example, on the typical Western diet, where 65–70 percent of daily calories come from carbohydrates, the first mouthful of food will likely contain carbohydrates. One of the enzymes these complex carbohydrates come into contact with is salivary amylase. It is produced by the salivary glands and begins digesting carbohydrates while they are still in the oral, laryngeal, and pharyngeal compartments seconds or even minutes before they reach the stomach. This makes it easier on the digestive enzymes produced by the gut and pancreas (e.g., pancreatic amylases) to which these dietary carbohydrates will subsequently be exposed.

Once dietary carbohydrates reach the stomach and duodenum (the first segment of the small intestine), they stimulate, from the walls of these tissues, the release of digestive products including hormones. Several such products, including pancreazymin and cholecystokinin, are released into the circulation, where they get delivered to the pancreas and liver to stimulate production and release of pancreatic amylases and liver enzymes. Other released products remain confined to the lumen of the stomach and gut, where they participate directly in the digestion and absorption of a meal.

Perhaps the best-known of these digestive products is gastric acid (or, more correctly, hydrochloric acid, HCl). A physiological solution with a neutral acid/base content, gastric acid has a pH of about

7.4 (0 being the most acidic and 14 the most basic). By the time a
meal is well mixed in the stomach, its pH is less than or equal to 2.0.
This increased acid or hydrogen ion content enhances the digestion
of ingested food products such as the protein in meat. Whenever
the gastric pH drops below 3.0, pepsinogens (the inactive form of
protein-digesting enzymes) are rapidly converted to pepsins (the
active form) that break down the protein in a meal.

The concentration of hydrogen ions (acid) in the stomach is
state dependent. In the basal state (i.e., immediately before ingestion of a meal), hydrogen ion concentration is at its zenith, and
pH is at its lowest level. As a meal is ingested and the stomach fills,
the hydrogen ion concentration decreases, and the pH rises. This
is because the ingested food acts as an acid buffer in the stomach.
Therefore, 60–90 minutes after the meal is consumed, stomach pH
is at its highest value, and gastric acid is at its lowest concentration.
Soon thereafter, however, as the stomach empties and the gastric pits
increase their production of hydrogen ions (the activated state), the
pH of the stomach contents declines, and the hydrogen ion concentration increases.

Transition from the basal to the activated state occurs because
sensory receptors (sensors) detect changes as the organ fills: chemoreceptors detect changes in the chemical composition of gastric
chyme, mechanoreceptors are sensitive to changes in stretch or dis-
tention of the wall as the stomach fills, and osmoreceptors respond
to changes in the osmolarity of the contents of the stomach. Mechanoreceptors are either smooth muscle cells, nerves, or other excitable cells. In the stomach, most of the innervation comes from the
vagus nerve.

As the stomach fills and stretches, the vagus nerve sends sensory
impulses to the central nervous system. This results in reflex motor
output to the stomach and other segments of the gastrointestinal

tract (GIT). The motor activity is carried by the vagus nerve to the corpus (body) and antrum (distal portion connecting to the duodenum, the first segment of the small intestine). In the corpus and antrum, acetylcholine—one of three important gastric acid secretagogues—interacts with delta cells (D cells), enterochromaffin-like (ECL) cells, and gastrin-producing (G) cells.

In the absence of vagal influences, D cells release somatostatin, which inhibits release of histamine from ECL cells. In the presence of acetylcholine, release of somatostatin by D cells is inhibited. Therefore, somatostatin cannot block release of histamine by ECL cells, and secretion of histamine is enhanced. The stimulated vagus motor nerves also release a neurotransmitter called *gastrin-releasing peptide* (GRP). It acts on G cells to increase the production and secretion of gastrin. Gastrin directly stimulates gastric parietal cells to produce gastric acid.

Secretagogues: Acetylcholine, Gastrin, and Histamine

Any compound that stimulates release of acid by the stomach is called an *acid secretagogue*. Of course, the term can apply to the release of compounds by other tissues and organs as well. There are three main physiological and naturally occurring secretagogues that stimulate stomach parietal cells. They are acetylcholine, gastrin, and histamine. Each can act directly or indirectly. When acting directly, they arrive at a parietal cell and bind to a distinct receptor that recognizes them. The parietal cell will then increase its production and release of hydrogen ions. When acting indirectly, acetylcholine and gastrin cause acid secretion by stimulating the release of histamine from ECL cells.

Inhibition of acid secretion is accomplished by somatostatin. Somatostatin is a polypeptide hormone made by D cells that are located in the antrum and corpus of the stomach. Somatostatin is also manufactured by D cells in the pancreas and by neurons in the hypothalamus. Somatostatin inhibits acid release by both direct and indirect mechanisms. In the direct pathway, it binds to receptors on the parietal cell and inhibits acid production by impairing function of adenylyl cyclase (an enzyme that produces cyclic nucleotides). Somatostatin's net effect is to antagonize histamine's stimulatory actions, which augment release of acid. Indirectly, somatostatin inhibits release of histamine and gastrin by ECL cells and G cells, respectively.

MECHANICAL ACTIVITY OF THE
GASTROINTESTINAL TRACT (GIT)

Of course, the first mechanical activity of the GIT takes place in the mouth. Once we have taken a mouthful of food, we begin to chew it (mastication). This breaks food into smaller pieces and simultaneously mixes it with salivary secretions. Both help in the downstream mechanisms of digestion and absorption. It goes without saying that most people chew food with their teeth. We have three specifically designed kinds of teeth: incisors, canines, and premolars/molars. As a rule, in the healthy adult, there are four top and four bottom incisors. From the Latin, *incise* means "to cut." Our eight incisors, with their flat, sharp edges, were designed to cut food (not fingernails). The four canine teeth (two above and two below) are for tearing and shredding. They are especially effective when meals include steak or other meats.

In nature, the canine teeth of carnivorous predators also serve to subdue their prey. Visualize a pride of female lionesses or a large male lion catching an antelope (or a taunting hyena). Lions and other large feline predators, if given the choice, go for the throats of their prey. Their canine teeth puncture large vessels that carry significant volumes of blood (e.g., carotid arteries and external/internal jugular veins). Their powerful jaws crush or otherwise occlude the trachea. In less than a minute, their prey will be deprived of blood flow to the brain, the return of venous blood to the heart, and airflow to the lungs. The animal is rendered asphyxic, hypoxic, and hypovolemic; these acute pathophysiological conditions result in the simultaneous collapse of the circulatory and respiratory systems and death.

In addition to incisors and canines, the majority of human dentition is composed of premolars and molars. More than anything else, molars and premolars are needed to crush food (e.g., grains, nuts, and seeds). Our teeth have been constructed to partially process almost any food that can be taken into the mouth.

Once a bolus of food has been pinched off and a mouthful (caused by an elevated tongue pressing food against the hard palate) is prepared for swallowing, the real intelligence of the GIT system unfolds. Arguably, it begins with the swallowing reflex. The swallowing reflex includes a large number of constituent precision reflexes designed to pass the bolus from the mouth into the esophagus. Simultaneously, other portals—most notably the trachea and nasal cavity—are closed. As the upper esophageal sphincter relaxes reflexively to permit entry of the bolus into the esophagus, the epiglottis is drawn down and over the open glottis (also reflexively), ensuring that the food does not enter the trachea, which could lead to aspiration, choking, and suffocation.

When the bolus is safely inside the upper esophagus, a smartly ordered sequence of events takes place that ensures the downstream delivery of the food to the stomach. First, the upper esophageal sphincter contracts, closing its proximal end. This prevents regurgitation of the bolus and potential aspiration into the trachea. Second, an esophageal peristaltic wave is initiated. This is characterized by the contraction of muscles immediately upstream to the bolus and relaxation of similar muscles downstream to the bolus. This complementary action ensures that a luminal pressure gradient (high pressure upstream and low pressure downstream) entraps the bolus of food and moves it in an oral-to-aboral direction.

As a side note, the human esophagus can be divided into three segments. The upper third is composed mostly of striated skeletal muscle. The middle third is a mixture of diminishing skeletal muscle and increasing visceral smooth muscle. The final third is almost entirely visceral smooth muscle (as is the wall of the rest of the GIT system down to the rectum and anus).

As the esophageal peristaltic wave moves the bolus toward the lower esophageal sphincter, another constituent of the swallowing reflex relaxes both the lower esophageal sphincter and the stomach. This is referred to as *receptive relaxation*. Receptive relaxation allows passage of the bolus through the lower sphincter and into the stomach. Moreover, receptive relaxation prepares the stomach to accommodate a large volume of solid and liquid materials that compose any particular meal. As the stomach relaxes, its compliance increases. This means that under such conditions, the stomach can accommodate large volumes of undigested materials without developing unusually high pressures.

Sphincters in the gastrointestinal tract are regions of densely concentrated visceral smooth muscle cells that are circumferentially arranged at the beginnings and ends of visceral organs. Five or six important sphincters help characterize the GIT system. Actually, there are more than this if we include the ductal sphincters of the pancreas, gallbladder, and liver. For the moment, however, I am ignoring the latter. In an oral-to-aboral direction, the important sphincters of the GIT system are as follows: (1) the upper esophageal sphincter, (2) the lower esophageal sphincter, (3) the pyloric sphincter, (4) the ileocecal sphincter, (5) the internal anal sphincter, and (6) the external anal sphincter. I will treat (5) and (6) as if they were one.

When contracted or closed, any two adjacent sphincters define an anatomically separate compartment of the GIT system. Each compartment has distinct functions. For example, the lower esophageal sphincter and the pyloric sphincter help define the stomach, or *gastric compartment*. Shortly after ingestion of a meal, contractions and movements of the stomach wall (caused by gastric smooth muscle) churn and mix the stomach's contents, producing a semiliquid suspension called *chyme*. The churning and mixing ensure that all components of the meal—carbohydrates, fats, proteins, and water—are efficiently exposed to gastric and salivary secretory products, including salt solutions, acids, and enzymes.

Another compartment is established by simultaneous closure of the pyloric sphincter (the distal segment of the stomach where it joins with the duodenum) and ileocecal sphincter (the junction of the ileum, or *distal small intestine*, and the cecum, or *ascending colon*). This is the small intestine. Each of the three segments in this compartment is characterized by distinct types of cells and corresponding differences in digestive and absorptive functions. For example, after preparation in the stomach, the bulk of every meal is digested and absorbed in the duodenum or the first segment of the small intestine. This means that the enzymes needed to break down carbohydrates, fats, and proteins are either found within the lumen of the duodenum or they are constituent elements of the cell walls that make up the mucosal lining of the gut.

The winter of 2016 and 2017 was unusually mild in the northeastern United States. In our yard, we had flowers budding and blooming in February. I think that both plants and animals were confused by the unexpected mild weather. As predicted, the late winter and early spring of 2017 were unusually potent for allergens, especially the airborne varieties. Many friends and family members, including my wife and me, experienced early allergy symptoms and some previously inexperienced effects. For example, there seemed to be viruses everywhere, and many of us caught colds and the flu. My wife had repeat bouts of these illnesses between early March and mid-May.

In late April 2017, she and I attended the annual convention of Experimental Biology in Chicago. I still had my cough from the end of March and had been medicating myself for about a week prior. I was taking DayQuil and NyQuil gelcaps twice daily. These over-the-counter medications contain a variety of compounds: dextromethorphan, a cough suppressant; acetaminophen, for pain and fever reduction; phenylephrine and ephedrine, to contract smooth muscle and affect blood flow and blood pressure; doxylamine succinate, a histamine blocker; and so on. All of these compounds have side effects. Once I experienced relief from the cough suppressant, I failed to give further consideration to the other ingredients.

We arrived in Chicago on a Friday evening. On Saturday evening, we were having dinner when I experienced difficulty swallowing my food. For the drier contents in my mouth, I had to drink water to activate the swallowing reflex and get the food down. This made me nervous, and it persisted for the next several days while in Chicago. I was confused about the potential causes. I wondered if this was simply another symptom of my earlier cold but having a late onset. I was also taking a few thousand milligrams of vitamin C each day and wondered about that. I wondered if the difficulty could be caused by a single ingredient in the DayQuil or NyQuil or by a combination of ingredients.

Knowing a good deal about the physiology of the swallowing reflex, it seemed to me that my upper esophageal sphincter was not operating properly. Instead of relaxing on the first attempt at swallowing, it took several attempts, my head tilted backward to extend my throat, and drinking water to get the meal down. Also, instead of taking 30 minutes or so to eat a meal, it now took me 45–60 minutes. This was inconvenient, uncomfortable, and worrisome. My poor wife had to endure my behavior with great patience.

With much thought and after a couple days in Chicago, I decided to discontinue taking DayQuil, NyQuil, and any other medication (e.g., vitamin C and an occasional aspirin when my diet included too much fat). Within a week of discontinuing all medications, my ability to swallow began to return. After two weeks, it was fully restored. I concluded that dextromethorphan, ephedrine, phenylephrine, doxylamine, a combination of these, or some other cause (such as the illness I was still combating) had partially paralyzed the neuromuscular apparatus that operates my upper esophageal sphincter.

CHYME AND MECHANICAL BEHAVIOR OF THE GIT

Once chyme enters the second compartment described above, the intelligent intestine knows exactly how to either digest it further or immediately process it for absorption. Considering the individual molecules in a meal, there are several possible mechanisms for a meal's digestion and absorption inside the small intestine. For example, one particular molecule might need no further digestion and is immediately ready for absorption across the wall of the gut and into the circulation. Glucose, or blood sugar, is one such example. Glucose is a small, simple monosaccharide that needs no digestion once inside the stomach and small intestine. It only needs to identify the cell-bound protein carrier to which it must bind. Once bound, glucose is transported into the mucosal cell, either for use or for delivery across the basal membrane and into the interstitium and circulation. Because of concentration gradients, glucose can also diffuse between two adjacent cells before it enters the systemic circulation.

Conversely, some molecules are so large and complex that once inside the small intestine, they must be exposed to luminal enzymes for reduction into smaller compounds. Subsequently, they are further exposed to enzymes that are attached to the luminal side of mucosal cells for additional digestion (complex carbohydrates such as pectin and related compounds). Only then are they small enough to cross into the circulation. There are yet other, more complex dietary molecules that even after luminal and mucosal digestion must be reassembled inside the cell before being transported not into the systemic circulation but into the lymphatic circulation because they are too large to cross systemic circulatory barriers. Dietary triglycerides and other integral components of micelles illustrate this pathway of digestion and absorption. Micelles are geometric cylinders composed of free cholesterol, other cholesterol-related compounds, free fatty acids, and complex dietary lipids conjugated to bile acids and salts (derived from the liver and gallbladder). Micelles and chylomicrons—vehicles for the circulatory transport of cholesterol, high-density and low-density lipoproteins (HDL and LDL, respectively), as well as triglycerides—represent the most complex and difficult digestive and absorptive processes known to physiology.

Additional important components of the mechanical behavior of the GIT system are the reflexes that move chyme en masse from one location to another. For example, powerful contractions of the stomach antrum can move gastric chyme in a retrograde direction toward the esophagus. When such retropulsion waves move in the retrograde direction, they are often confronted by antegrade contractions moving food in the opposite direction. These head-on collisions of directionally opposite gastric contractions cause favorable mixing of chyme, thereby maximizing exposure of its contents to the products of salivary and gastric secretions.

The large intestine is equipped to store the residual contents of several meals and to extract as much remaining water as possible. It also serves to house commensal bacteria, which act symbiotically with the large intestine. These bacteria get a comfortable dorm to live in, and they provide beneficial functions to us by detoxifying gram-negative bacteria and producing gases that help move our bowels. The colon is characterized by several varieties of contraction. These include haustration, propulsion, and segmentation. Individually and collectively, such mechanical properties serve to extract remaining nutrients and water and to minimize storage time of dietary waste products. Further speaking in generalities, these colonic movements are confined to local regions extending no more than several inches. On occasion, however, high-amplitude, long-distance migrating contractions occur. These are referred to as *high-amplitude* and *migrating* because of the unusually high bowel pressures they generate and because they are known to move over great lengths of the large intestine (e.g., 15–18 inches or more).

To help diagnose bowel disease, clinicians try to replicate what actually happens in the large intestine. In the case of faulty bowel mechanics, clinicians investigate by inserting a double-lumen catheter into the patient's empty rectum. The catheter is advanced up the descending, across the transverse, and into the descending colon (the cecum). One lumen of the catheter is attached to a flaccid inflatable balloon. The other lumen records pressures inside the colon. The clinicians then fill the inflatable balloon with warm water, warm saline solution, or air and simultaneously measure luminal pressure and contractions. The filled balloon is designed to replicate a colon filled with fecal matter. From such exercises, investigators have found that high-amplitude migrating colonic contractions can generate pressures that exceed 100 mmHg. If the filled balloon fails to generate these high pressures, then the physician suspects neuromuscular dysfunction of the GIT and further tests are performed to identify the problem.

In a healthy colon filled with feces, high-amplitude migrating contractions can expel stools in excess of one pound in weight. Such contractions and corresponding bowel movements can come on suddenly and with little warning. When they do, it is important to be as near a restroom as possible.

HOMEOSTASIS AND BOWEL FUNCTION

In the average healthy adult whose life is routine for at least 8–12 weeks, gastrointestinal behavior is also routine. To me, *routine* means that sleep habits do not vary. We go to bed and arise at the same times each day. Dietary habits remain constant. For example, we might regularly eat three meals of similar compositions at the same times each day (i.e., breakfast, lunch, and dinner). Work and leisure times are fixed, and one's general comings and goings are consistent (e.g., no periodic or intermittent international travel). Anything other than this arguably represents the nonroutine. When lifestyle is routine, the likelihood of one's bowel functions being in a state of homeostasis is high. If your life is routine and consistent with the above, and if your bowel movements occur daily in the morning hours, then you can probably conclude that your GIT function is in the steady state and healthy.

Now imagine that your employer assigns you to work in Asia or in North or South America for a two-week period and that you must leave in less than one week. There is a high probability that all of the above will change—dietary practices, sleeping habits, daily comings and goings, and so on. Even if you can control most of the variables I have described as being routine, you cannot influence the difference in time zones to which you will be exposed during such a foreign assignment. More than likely, when you return home three weeks later, you will notice a marked change in GIT function. The earliest indication could well be the irregularity of your bowel movements. Instead of occurring daily as they did before the international travel, you might go several days or even a week without a bowel movement. Also, you will probably notice a change in the volume and consistency of your stools. These and other changes will confirm the travel-related disruption of physiological GIT function. It can take weeks or even a month or more to reestablish your pretravel GIT routines.

A couple personal experiences might help illustrate this point. My wife and I went for a walk one Saturday morning. At an intersection, we had to stop because she felt pain in her chest and was out of breath. For several minutes she was immobilized. We were 10–12 minutes from our house, and it took more than that to return. We were not sure what was happening but considered heart attack and pleurisy as two possibilities. At home she took an 81-milligram dose of aspirin and 1,000 milligrams of acetaminophen (based on my research) as I rushed her to the nearest hospital emergency room. Blood, urine, and other tests were performed, and a heart attack was ruled out even though she still had periodic chest pain and difficulty breathing. Our physician and the examining cardiologists decided to admit her for overnight observation. Her condition improved, and she was released the next day.

Before leaving the hospital, my wife was prescribed cholchicine by the cardiologist and began taking it the second day after her event. Cholchicine is used mainly to remedy gout and acute attacks of arthritis. However, it has also been prescribed to treat pericarditis (inflammation of the pericardium secondary to a heart attack or to viral infection). Although he didn't say so, these were probably the reasons for the cardiologist's prescription. Within two or three days, the regularity of my wife's GIT system was in disarray. Because the physicians had told us the lab tests revealed no signs of heart attack, my wife discontinued the cholchicine after several more days. However, her bowel regularity did not return for several months.

In the meantime, we learned that cholchicine is extremely toxic in high doses and that the GIT system is one of its main targets. We already knew from years of observation that my wife has a sluggish cardiovascular system. This is evidenced by her persistently low blood pressure, minimal responses to stimuli that regularly increase heart rate and blood flow, and fingers and toes that are almost always cold except in the heat of July and August. This change in routine (taking a toxin, cholchicine, to treat a condition that probably did not exist) was as disruptive to her GIT system as three weeks of unexpected foreign travel can be.

When our youngest son graduated from the marine corps boot camp at Parris Island, South Carolina, my wife and I attended the ceremony. A few days later, she and I had dinner in Charleston, at a Mexican restaurant that came highly recommended. My dish included refried beans. Dinner was at about 7 p.m., and by midnight, I woke up violently ill with food poisoning. At that moment, and each hour or two thereafter, I threw up and had diarrhea. This continued until about 8 a.m. As we drove back to New Jersey, I was powerless, dehydrated, and malnourished. I didn't dare eat or drink anything for the next 24 hours and then went on a strict BRAT diet (bananas, rice, applesauce, and toast without butter or other toppings). After this unexpected change in routine, and unavoidably, my GIT functions and bowel regularity did not return for six to eight weeks.

For many years before these incidents, I had observed the effects of breaking routine on my own bowel movements. I traveled to scientific conferences much more frequently in my 30s, 40s, and 50s than I have since. Those trips usually lasted four to five days except when they included international travel. My sleeping, eating, and daily comings and goings were anything but routine. All normalcy for the life of my GIT system was predictably upset by such travel, and I knew it before, during, and after the conferences for several weeks. Both our bowels and our bowel functions like homeostasis and routine. They do not like disturbances.

Our Intelligent Gastrointestinal Reflexes

To me, all body reflexes show intelligence. To develop this point, let's recall what a reflex is by reidentifying its component parts (see *Our Marvelous Bodies*, Rutgers University Press, 2008). In the human body, all reflexes have five physical components. They are (1) sensory receptors (also called detectors, sensors, etc.); (2) afferent sensory nerves (also called afferent nerves, afferents, etc.) that connect sensory receptors to the third component; (3) central integrators, which are collections of neurons found in the spinal cord, the brainstem, and/or higher brain structures and that integrate and interpret incoming messages from afferent sensory nerves and transmit an intelligent response via the fourth component of the reflex; (4) efferent motor nerves (also called efferents or motor nerves), which convey the integrator's messages; and (5) activators or effectors that are peripheral to the central nervous system and receive these messages. Activators and effectors are often skeletal and/or other muscle types, secretory endocrine or GIT organs, or multiple other tissues and organs.

The GIT system has its own network of nerves called the *enteric nervous system*, or the "little brain." Reflexes involving the enteric nervous system often, but not always, originate in the lumen or wall of the gut (the sensors) and convey their messages back to the same or to other compartments of the GIT and their constituent activators (e.g., visceral smooth muscle cells, mucosal secretory cells lining the walls of crypts, and pits that might serve as local autocrine organs). In other cases, the sensory input comes from the peripheral nervous system and is transmitted to the GIT by the autonomic nervous system via both the parasympathetic and sympathetic branches of this efferent motor system.

Gastrointestinal reflexes are activated physiologically by two sets of inputs: mechanical and chemical. Mechanical stimuli arise from changes in volume of any given compartment that are caused by the filling or emptying of that compartment. In addition, the chemical composition of the chyme occupying a compartment provides another source of stimulus to our GIT reflexes. For example, it is well established that a given volume of a liquefied meal will exit the stomach much faster than a mostly solid meal composed of large particles. Moreover, the rate of emptying of the stomach is faster for a meal made up mostly of carbohydrates than it is for a fat-laden meal. Additionally, meals of greater osmolarity are emptied by the stomach slower than are meals of lesser osmolarity.

All GIT function can be characterized by phases. These include the cephalic, gastric, and intestinal phases. *Phase* refers to the period of time during which one or more GIT functions are affected by an activity or event. The cephalic phase occurs when food is being prepared. The sights, smells, and sounds associated with the preparation of a meal have potent effects on our GIT systems. They stimulate salivation and secretion and help prepare the system for consumption of a meal.

Once we begin ingesting a meal, the gastric phase of digestion and absorption begins. This means that the various acts of chewing and swallowing help prepare the stomach to receive a bolus of food and subsequently the entire meal. For one thing, as a bolus of food is pinched from a mouthful and pushed toward the larynx by the elevated posterior tongue, the upper esophageal sphincter relaxes so the food can enter the esophagus. Near-simultaneously, the epiglottis closes entry to the trachea so the bolus is not aspirated and will not choke us. The stomach also relaxes by other reflexes (e.g., receptive relaxation), and its compliance increases so it can accommodate a meal.

After some time in the stomach, and as the stomach begins emptying its contents into the duodenum, the third, or intestinal, phase of digestion begins. This is the phase during which multiple enzymes are being prepared to come in contact with the carbohydrate, lipid, and protein contents of the recently ingested meal. These enzymes can be released into the lumen of the gut, where they participate in digesting and reducing the size of ingested molecules. Or they form an integral part of the mucosal membrane, where they will help further reduce molecular sizes, transport food products into cells, and reassemble products once they are inside the cell.

A couple examples will help solidify some of the information just presented. First, not all GIT reflexes have been identified. Those reflexes that have been published simply reflect the historical and contemporary work of investigators interested in the reflex. Considering the full cecum in the absence of disease, it is intuitive that communication between this compartment of the GIT and upstream compartments exists. That is, if the cecum is filled to capacity and any upstream compartment takes on an additional meal, then the cecum and downstream colon will need to empty before the colon can accommodate an additional volume. The ileocecal and other GIT reflexes are designed to accomplish this.

Unwisely trying to swallow too large a bolus, or swallowing before food has adequately been reduced in size by proper chewing, can and will lead to danger. It is all too easy to choke when doing these and similarly foolish things like talking excitedly while chewing a mouthful of food. My wife and I and one of our sons were hosting visitors in our family room one Saturday afternoon. A young girl ran across our lawn, rang the doorbell, and asked if Mr. Merrill could come quickly. I left our conversation and followed her through the neighborhood to an adjacent street. There was a middle-aged woman kneeling down and obviously choking. Only a couple mothers and a handful of children were nearby. One of the mothers had sent her daughter to find me. The other mothers were rounding up their startled children and taking them indoors.

I knelt beside the woman and asked if she could breathe. She shook her head no. I knew I had to get her into a position to perform the Heimlich maneuver, but her body was too large and flaccid for me to move alone. By the time I had made several unsuccessful attempts, she had collapsed but was still trying to breathe, and I could palpate her pulse—thus, I knew she was alive and was not in cardiac arrest. About this time, one of our township police officers arrived. He ordered me out of the way and retrieved his defibrillator from the car. As he was fumbling with it, I told him she was not in cardiac arrest and that it was unwise for him to try to apply the instrument.

Fortunately for her, the EMTs arrived a minute or two later. They were trained, well-equipped, and experienced. Two of them went to work, and within another few minutes, they had extracted a large bolus of unchewed chicken from her throat. She regained consciousness and began breathing deeply and frequently as the EMTs put her on a stretcher, placed her inside the ambulance, and sped off. Several days later, this neighbor came to thank me for my efforts and told us the story of her accidental choking. We were relieved that she was alive and in good health.

While not everyone who swallows a large bolus of food nearly chokes to death, many of us overeat and tax the anatomical and physiological capacities of our GITs. Imagine that you overate every meal over the course of several days without having a bowel movement. This is not an unlikely scenario for business travelers or those who change time zones, overeat, underexercise, and get constipated during such travel. These experiences reliably interfere with normal physiological function, including the activities of GIT reflexes that involve digestion, absorption, and mechanical movement of intestinal/colonic contents.

Chemical, neurogenic, and mechanical stimuli (e.g., stretch) in the stomach and the small intestine normally result in reflexes between these and downstream compartments such as the colon. Such reflexes lead to timely and routine emptying of the colon through healthy bowel movements. When daily or otherwise routine bowel movements are disturbed, one can go several days to a week or more without a bowel movement. This is unhealthy. After such prolonged periods, "housekeeping" bowel movements are not uncommon. That is, a peristaltic wave of contraction might move over a distance of 15–18 inches (38–56 centimeters) in the large intestine. Stored fecal matter in an equivalent length of the downstream large intestine (e.g., the transverse colon, descending colon, sigmoid colon, and rectum) is simultaneously eliminated. Such peristaltic waves are called *high-amplitude migrating colonic contractions*. As mentioned above, they can eliminate stools weighing in excess of one pound (454 grams).

The gastro-colonic reflex is one of a number of physiological reflexes controlling motility of the gastrointestinal tract. It involves an increase in motility of the colon in response to stretch in the stomach and/or to the presence of by-products of digestion in the small intestine. This reflex is responsible for the urge to defecate following a meal. Reflex communication between the stomach and the colon has been demonstrated in animals and humans by several methods, including autoradiography. In the latter case, a meal containing a contrast medium is ingested. After a few hours, the contrast medium arrives in the colon, when the stomach is empty. The stomach is then intubated with an inflatable balloon. As the balloon is inflated it stretches the stomach, initiating the gastro-colonic reflex. Monitors placed over the colon can record movements of the previously installed contrast medium to confirm the reflex.

The gastro-colonic reflex has also been investigated by myoelectric recordings in the colons of animals and humans. These records show an increase in electrical activity within the colon in as few as 15 minutes after eating. Such recordings also demonstrate that the gastro-colonic reflex is uneven in its distribution throughout the colon. In terms of a phasic response (here, periodic contractions), the sigmoid colon is more greatly affected than the ascending or transverse colons. However, the tonic response (here, sustained contractions) across the colon is less predictable. Moreover, when pressure inside the rectum increases, the gastro-colonic reflex acts as a stimulus for defecation.

Clinically, the gastro-colonic reflex has been implicated in the pathogenesis of irritable bowel syndrome. The very act of eating or drinking can provoke an overreaction of the gastro-colonic response in such patients due to heightened visceral sensitivity. This can lead to abdominal pain, diarrhea, or constipation. Certain medications, including antibiotics, can decrease the tonic responses of the stomach and gut to stretch and can affect both the frequency and quality of bowel movements—that is, these medications can be either constipating or loosening. When being prescribed medications, the intelligent patient should question the physician about the negative side effects, including those on the GIT system.

In terms of the onset of action, the fastest GIT reflexes are those that occur neurogenically. This is mainly because it takes less time for action potentials to travel over sensory and motor neurons than it does to synthesize proteins (minutes to hours, or longer) and to activate other secretory events resulting from chemical stimulation. Also, in the latter case, many chemical stimulants in the meal or in the wall of the GIT must enter the systemic circulation to be distributed to their target organs. The rate of delivery to such targets will then be dependent, in part, on the active or inactive blood flow to that tissue, whether the chemical stimulant gets bound to circulating proteins as carriers (and therefore must be released at the target), and all the factors that can influence this route of delivery.

BALANCE AND THE INTELLIGENT BODY

There seem to be more fad diets cropping up each year than there are fruits and grains growing on the earth. Many of these are found in books and journals that have been written by physicians, or they are promoted on television talk shows hosted by physicians and celebrities. Sorting out fact from fiction in these claims can be an almost impossible task for even the most intelligent among us. Fad diets are like gym memberships. Anyone who subscribes to one thinks she or he is a world expert on the topic.

Similarly, almost anyone who has shed a pound or two following someone's famous diet, including those who reduce their personal consumption of large sugary drinks, is an expert on the topics of calories and weight loss. The average such person probably could not lead an intelligent conversation on the subject of "energy balance" any more than they understand the physiology of "body water balance." As a physiologist, I should emphasize the word *balance* as the key to intelligent, successful living. A person living a balanced life needs neither a gym membership nor a fad diet to maintain good health and a near-constant body weight. Argument to the contrary is simply unintelligent.

ABOUT THE AUTHOR

GARY F. MERRILL is a professor of cell biology and neuroscience at Rutgers University. He received a PhD in physiology from Michigan State University in 1975 and worked as a cardiovascular postdoctoral fellow at the Louisiana State Medical Center in New Orleans from 1975 to 1976. His other works include *Our Marvelous Bodies* (2008) and *Our Aging Bodies* (2015), both published by Rutgers University Press. When he is not working on his writing, he enjoys hiking and camping and spending time with his children and grandchildren.

.

Printed in the United States
By Bookmasters